AGING RESEARCH

A Look at Some of the Scientific Evidence on Aging

Harold Massie, Ph.D.

Note for Librarians: A cataloguing record for this book is available from Library and Archives Canada at www.collectionscanada.ca/amicus/index-e.html
ISBN 1-4251-0141-0

PUBLISHING™

Offices in Canada, USA, Ireland and UK

Book sales for North America and international:
Trafford Publishing, 6E–2333 Government St.,
Victoria, BC V8T 4P4 CANADA
phone 250 383 6864 (toll-free 1 888 232 4444)
fax 250 383 6804; email to orders@trafford.com
Book sales in Europe:
Trafford Publishing (UK) Limited, 9 Park End Street, 2nd Floor
Oxford, UK OX1 1HH UNITED KINGDOM
phone 44 (0)1865 722 113 (local rate 0845 230 9601)
facsimile 44 (0)1865 722 868; info.uk@trafford.com
Order online at:
trafford.com/06-1898

10 9 8 7 6 5 4 3

This book is dedicated to Catherine Dery, the great grand-mother in our family. She is on the front cover at the age of 82 and on the back cover at the age of 16.

ACKNOWLEDGMENTS

Several people have contributed their editorial skills to this book. Deana Holmes, David Krom, and Hon. Debbie Massie have made large contributions to this effort. The grammar, syntax, spelling and general sense of the manuscript have all been greatly improved as a result of their efforts. Cover design is by Wakeman Massie.

Contents

INTRODUCTION

I became interested in the aging process as a result of living with my grandparents in Ohio for about six months when I was eleven years old. They were over 65 years of age and living on a Social Security check of $80 per month and not much else. Their rent was $25 per month for an old two story house which was heated by coal, and I remember my grandfather having a very lively discussion with the coal delivery man when the price went up to $10 a ton. He burned about seven tons for the winter. The water bill was 25 cents a month.

There was no government sponsored medical care in those days. Most older people just lived with their medical and economic problems and suffered. This seemed to be puzzling and unfair. So I wondered at a very early age why people grow old and how a strong man or woman could become so weak and dependent on others.

This book is not intended to be a scientific publication. I have, therefore, not referenced most of the material, almost all of which has been published in the scientific articles listed at the back of the book. It has taken more than thirty years to collect the information for this book.

Almost all of the information is stored in three filing cabinets and eight bankers boxes filled with reprints and data files. I have, however, referenced important and interesting texts within the manuscript for readers who wish to explore various issues.

The two main editors for the content of this book are my wife, Pat, and my daughter, Elizabeth. They both have little or no scientific background but have made a large contribution to the clarity and sense of this book. Much of the material has been changed, eliminated or clarified as a result of their efforts. Elizabeth has suggested that this book should be regarded as a detective story with the reader deciding what is pertinent and what is not. I think that she is right about this.

This is really a story about the efforts of a research scientist to understand the basic biological aging process. Some of the writing is a little technical (you could think of these sections as being forensic evidence), but every effort has been made to make the material clear to all readers. This is not a "how to do" book but rather a mystery story exploring the possible causes of human aging. Based on the evidence presented you can draw your own conclusions. The book also looks at how scientific research is done for those of you who wonder what those people in white coats really do.

I have frequently used the word "we" in this book because almost all of the experiments described in this manuscript were done with the help of other people, including research assistants, other scientists, postdoctoral fellows and summer students. Their names are included in the scientific references listed at the back of the book.

Chapter 1

DEFINING AGING

All living things age and grow old and finally die. Some plants, such as the Sequoia redwood and Bristle Cone pine trees, live for thousands of years while some viruses may live for only a matter of minutes. No matter how long any organism lives it will experience a decline in biological performance with time. We humans see this as a general loss of energy, strength and stamina. In addition, many changes occur which are of a more or less cosmetic nature, such as skin wrinkles, gray hair, hair loss, hair growing where it never grew before, weight gain, diminished vision and hearing, loss of teeth, sagging muscles and loose skin, just to name a few. No one really likes these things to happen, but they are not life threatening and are partly correctable if you have the time and money to deal with them. More than anything else, they reflect the ongoing process of human aging.

The real problem of human aging is the increase of serious and debilitating disease states. Heart disease, cancer and stroke are the three leading causes of death in the United States, and all of these diseases show definite age-related increases. Unfortunately, these conditions also occur in children and young people but, thankfully, only at a very low incidence. Ninety five percent of all the cases of these illnesses occur in people over the age of 40 with the incidence doubling with every seven additional years of age in both men and women. The same is true for every other country in the world where the average person lives to be more than 40 years of age.

It has been said that aging is a process which increases the probability of death from virtually all causes including, of all things, being run over by a car. The death toll from all this is very high with over 2,000,000 people perishing every year in the United States alone. Understanding this process would seem to be of utmost interest.

The principal approach to the problem has been to study the diseases associated with the aging process rather than the process itself. The statisticians have shown us that this probably will not be fruitful since completely curing the two major diseases associated with aging, cancer and cardiovascular disease, will add only a total of nine years to the average human life span. It is hard to believe that eliminating all forms of cancer will add only two years to the average life span. Curing all cardiovascular diseases nets an average increase of only seven years. That too, is an incredible but true statement. The aging process seems to indicate that if one disease or accident does not get you, then something else will.

Perhaps human aging is a natural and unavoidable phenomenon due simply to wear and tear or maybe there is a basic process that we do not understand, a sort of biological clock determining the length of human life. Even if it is simply wear and tear, what might we do to slow the process down? To understand the biological clock concept would also be an important approach. Thus, there are only two viable approaches to addressing the problem of human aging. The next chapter is devoted to these questions.

Chapter 2

MODEL SYSTEMS FOR STUDYING AGING

When I first began research on aging I was given the opportunity to visit a medium-sized pharmaceutical company research laboratory to talk with its director. He was an articulate man who seemed genuinely interested in aging research and how it might profit his company. I was given a tour of the research facilities and taken out for lunch. He took the time to offer me coffee and a very serious discussion. Basically he said that, if you have a drug or treatment for aging, you have to prove that it will work in humans and that it will be nontoxic. He said: "How will you be able to do this considering how long humans live?" I thought that a model system that ages rapidly was necessary. Model systems, however, by their very nature are inadequate. These model systems can offer only possibilities but not the real answers to the problem of human aging.

Many different investigators have used colonies of rats and mice to study the aging process. These colonies are very expensive to maintain. In general, both of these rodents live at most 2 to 3 years, which is a long time to wait for an answer to your experimental questions. The great advantage of this system is that scientists are dealing with mammals which means that what is found may also be expected to occur in humans.

Another approach to aging research involves the use of human cells in tissue culture. Cells can be grown in plastic bottles and studied under very demanding sterile conditions. The first of these cell lines was started by Leonard Hayflick who worked at the Wistar Institute and was called WI-38. These cells divide in bottles about 50 times and then fail to grow and die out. This seems to represent a model aging system with the advantage that these are honest to goodness human cells. My own experience has been that this is an expensive and time intensive system requiring highly trained and careful technical people. This system, however, has and will continue to lead us to a better understanding of the basic aging process.

Nematodes are microscopic unsegmented worms that are pests for agricultural plants, especially in the Gulf Coast areas of the United States with sandy soil. They live for about a month and have been used primarily for important genetic studies of the aging process. I have not worked with this system and cannot make meaningful comments about it, but certainly this economical option has provided and will continue to provide interesting information.

Then there are fruit flies. These are the little flies that you see hanging around bananas and grapes. They have big red eyes and, unlike their cousins the houseflies, do not carry disease. They are attracted to yeast and especially to alcohol and are, therefore, a pest in most wineries. I have seen them clustered around the top of a whiskey bottle in a state of stupor. Scientists call them Drosophila, which means that they love (phila) fruit (droso). Fruit flies are the most highly evolved of all the insects. At room temperature they live only 50 days. In the laboratory they live on a diet of mostly corn meal and molasses with other added food ingredients in small glass tubes and can be maintained at very low cost in constant temperature incubators.

A great deal of genetic research has been done for more than a hundred years using fruit flies; even behavioral mutants have been produced and identified. Numerous biochemical studies have also been published. Important aging research studies have been conducted as early as 1925. The effect of environmental temperature on life span has been well-studied using fruit flies. Drosophila produce no body heat. These studies have shown that controlling body heat by using incubators at various temperatures can change life span. The changes are very large with flies at low temperature (11 degrees Celsius, 51.5 degrees Fahrenheit) living 150 days while those at high temperature (30 degrees Celsius, 86 degrees Fahrenheit) live only 20 days. This is an incredible difference.

Strain (which refers to race in Drosophila) differences have also been shown to result in large differences in the average life span. Flies from the state of Oregon, which are called Oregon R, live almost twice as long as flies from Sweden referred to as Swedish C. Everything about them appears to be the same, and yet there is an enormous difference in the length of time that they live. Why would this occur? We, and others, have made an effort to answer this question.

For anyone looking for the basic reasons for aging, fruit flies offer a model system where very great changes are known to occur naturally. The problem is that they are, after all, insects and their aging process may have little or nothing to do with human aging. As one person said to me, "I do not want flies to live longer; I want to kill them." My reply was that I only want them to live longer so that we may find a way to delay what is killing most of us, aging.

You either believe in the research results with fruit flies and other organisms or you do not. If there is a connection to the human condition, this would be marvelous. I believe that there is such a connection. Since all organisms age and finally die, there is the real possibility that the basic process behind all of this is universal. The study of this process and the results of those studies may, therefore, be of relevance to all organisms, including humans.

Chapter 3

CHANGES IN ENVIRONMENTAL TEMPERATURE

It is worth repeating that fruit flies do not produce their own body heat. This is also true for most other insects and for reptiles. It is, therefore, easy to study the influence of body temperature on the aging process in fruit flies simply by controlling the environmental temperature. This is done by keeping the flies in incubators where the temperature, humidity and light cycles are held constant. The differences in life span at the various temperatures are really quite amazing. At 35 degrees Celsius, which is close to the human body temperature of 37 on the Celsius scale, Drosophila live on average only about 3 days. At 25 degrees Celsius (77 degrees Fahrenheit) they live about 50 days and at 11 degrees Celsius (51.5 degrees Fahrenheit) they live 150 days or more. These studies have been done numerous times by many investigators and there is no doubt of their validity.

It should be said that flies at low temperatures are not very active, but at high temperatures they are very active. Lowering the human body temperature, and consequently the activity level, would probably not be a desirable thing to do even if it were possible. It should be kept in mind that mammals have a totally different kind of metabolism from that of insects. Mammals maintain a constant body temperature, whereas insects do not.

Rate of living theory

The important idea to come out of this research is that metabolic rate, or the rate of living, as it was first called by Pearl in 1928 (The Rate of Living, Knopf, New York), can determine the rate of aging and consequently life span, at least in fruit flies. There is other evidence to support this idea. In 1961 it was reported by Kibler and Johnson (Journal of Gerontology, vol. 16, pp. 13-16) that a colony of rats kept at a room temperature of 83 degrees Fahrenheit (28 degrees Celsius) lived a lot longer than another colony kept at a chilly 48 degrees Fahrenheit (9 degrees Celsius). I have referenced this only because the results were so striking. At 550 days of age only 12% of the chilly group were still alive whereas an astonishing 92% of the rats kept at 83 degrees were still alive. Since rats do not wear sweaters and coats, they experience a definite heat loss at the lower temperatures. The only way they can maintain constant body temperature at low temperatures is to increase their metabolic rate by eating more food and burning more oxygen. The experimental result was, therefore, that increased metabolism and increased heat production resulted in a very large increase in the rate of aging.

The rate of living theory of aging says, basically, that we are all given a full tank of gas at birth, and when that gas is all burned up, we come to a stop. Some burn it faster than others. Professional athletes and laborers, for example, generally have shortened life spans. There are, of course, numerous exceptions to this observation. Women in general are smaller than men, but they also have more insulating subcutaneous fat under the skin and a lower metabolic rate than men. Currently, they live an average of 5.3 years longer than men. That is why they can buy life insurance at much lower rates than males of the same age.

Surface to volume ratio
Another piece of evidence supporting the "rate of living" theory is the concept of surface to volume ratio in mammals. The bigger a mammal is, the smaller this ratio is. It takes a lot less skin per pound to cover a whale than it does to cover a mouse on a per weight basis. Mice live at most two or three years, whereas fin whales are believed to live for hundreds of years. An elephant has been reported to have lived 98 years. The observation is that, the higher the surface to volume ratio, the shorter the life span. The explanation proposed is that this is due to the difference in the rate of heat loss though the skin. Mice lose a lot of heat per unit weight when compared to an elephant. The only way they can maintain a constant body temperature is to have a much higher rate of metabolism when compared to an elephant. No one, unfortunately, has had the nerve or the financial support to see if putting sweaters on mice would increase their life span. I'll bet that it would.

This observation of surface to volume differences applies only to mammals, but there are exceptions. Bats, for example, live much longer than other mammals of comparable size. Rats, too, are an exception; they are 10 to 20 times bigger than mice and yet have the same life span. Maybe they have less fur. Birds throw the whole concept of surface to volume ratio into question. They have a much higher body temperature (45 degrees Celsius) than mammals and yet live a lot longer than mammals of similar size. The owl, Bubo bubo, for example, lives to be

68 years of age. Birds have feathers and everyone knows that there is nothing warmer than a down comforter on a winter night. Perhaps insulation is the key to a low-tech approach to aging. It is also curious that both owls and bats are nocturnal creatures which will be the subject of another chapter on how light affects the aging process (Chapter 20).

Hibernation

Bears and other animals hibernate during the winter. Their metabolism drops to very low levels, just enough to maintain life. If the rate of living theory is correct, then hibernation should prolong life span. In fact, it does, at least in Turkish hamsters. The longer these hamsters spend in hibernation, the longer they live. The differences are very large with a life prolongation of more than 50%. This, of course, would not be of any practical use to most humans. It just offers more support for the rate of living theory.

Since 30% of the heat loss in humans is through the head, perhaps people who wear hats in the winter live longer. The overall conclusion here is that metabolic rate and heat loss seem to have a lot to do with biological aging.

Chapter 4

THE GENETIC BASIS OF AGING

There is no doubt that genetics play an important role in the aging process. Different species have greatly different life spans. An elephant certainly lives a lot longer than a fruit fly. Even within a species there can be big differences in average life span. As previously mentioned, the Oregon R strain of Drosophila lives more than 50 days at 25 degrees Celsius (77 degrees Fahrenheit), whereas the Swedish C strain lives only 30 days under identical conditions. Considerable research has been done on the genetic basis of aging in fruit flies. I have not been involved in this impressive research, and I am, therefore, not qualified to make meaningful comments.

Some laboratory strains of mice and rats live longer than others but not by very much. The kangaroo mouse is an exception with a life span of more than 7 years compared to laboratory mice that live only 2 to 3 years, but I know of no one who has a colony of these interesting mice.

In humans there appear to be some differences in life span for different races, but these differences are not great. Blacks in the United States live 6 years less than whites and the difference is even greater for Hispanics. The question is whether or not this represents true racial differences or differences in economic factors. The wealthy in the United States have the greatest life expectancy of any economic group regardless of other factors. Perhaps this is due to the fact that more than 45,000,000 Americans have no health insurance.

Currently, Japan has the greatest average life expectancy for any country, in spite of the fact that they have one of the world's highest rates of cigarette consumption with more than 34% of adults admitting to be smokers. Whether or not their great life span is because of relatively homogeneous racial makeup or to environmental and social factors is not clear. Most of the differences in human life span in the various countries of the world seem to be due primarily to economic and health care factors. Malnutrition and disease still remain the major factors affecting human life span in many of the countries of the world. At this date, 3,000 children die of malaria every day according to the World Health Organization. There are, however, two serious disease conditions which may represent examples of genetically based human aging.

Premature aging

One is progeria. Progeria is a rare disease characterized by all the symptoms of old age occurring during the teenage years. These kids get the whole works with wrinkles, loss of energy, chronic infection and cardiovascular disease. Fortunately, there are only about 50 cases known throughout the world. What impresses me is that these children have no subcutaneous fat. This layer of fat under the skin is a source of body insulation resulting in less heat loss. These children look very old, and they die before age 20.

Another is Werner's syndrome. This is also a rare disease where suddenly around the age of 30, a rapid rate of aging begins which soon results in death. Even less seems to be known about this condition than is known about progeria, except that the usual appearance of great old age and the disabilities and diseases of old age are as obvious as they are for progeria.

Fortunately, these two diseases are extremely rare, but they do offer an insight into the possible genetic basis for human aging. Is there something missing from the genetic material of these individuals and, if so, what or where?

Nuclear DNA

All genetic material resides in the long filaments of DNA (deoxyribonucleic acid). More than 95% of this DNA is coiled up in the chromosomes at the cell's center, the nucleus. The cells make up the organs of the body. Different cells work together to accomplish the task of a particular organ such as the kidney with everything being directed by the information coded in the DNA molecules. The language of this code has only four letters: A,T,G and C. The sequence of these four letters is the language of genetics. The DNA in the chromosomes of the cell really represents a library where almost all of the information needed to run a cell is kept.

All cells in the body contain the same amount and same kinds of chromosomal DNA, but only a small amount of it is used by a given cell type. Liver cells use completely different chromosomal DNA than brain cells do, even though they both have the same chromosomal DNA. This is all controlled by suppressor proteins attached to the chromosomal DNA. They control what kind of DNA a particular cell uses. If these proteins fail, then the possibility exists for liver information to appear in a brain cell or a kidney cell. We do not know if this sort of thing occurs with aging, but it certainly seems possible.

Damage to the very long DNA molecules coiled up in the chromosomes seems to occur with aging. In rats it has been shown that some breakage of the molecules occurs with increasing age. Massive loss, however, of the A, T, G or C code molecules does not appear to happen. The same is true in fruit flies. Overall, there seems to be very little damage to chromosomal DNA. This is probably because all sorts of repair processes exist in the nucleus of the cell to deal with broken DNA molecules and other forms of damage to the DNA. Repair enzymes accomplish this. Enzymes are simply proteins that work for a living. They do all sorts of marvelous things within the cell in contrast to their cousins, the structural proteins, which make up tendons, the matrix for bones, skin and other organs.

Mitochondrial DNA
While 95% of the DNA is in the nucleus in the form of chromosomes and is responsible for a lot of things that might go wrong, the other 5% is in the mitochondria. These are small organelles located in the cytoplasm surrounding the nucleus. There can be hundreds or even tens of thousands of these mitochondria in all cells. They are about the size of bacteria and are surrounded by a membrane barrier. Some scientists have even suggested that they are the descendents of ancient bacteria that infected cells a long time ago. I doubt this theory since bacteria have DNA repair enzymes, whereas mitochondria do not. Mitochondria are literally miniature power plants. They produce energy and heat.

Many investigators have reported a decrease with aging in the number of mitochondria in tissues from several different sources, including humans. In mice the number of mitochondria in brain cells decreases with age and in humans the number of mitochondria in liver cells decreases with age. This probably occurs elsewhere as well, but the studies have not been done.

All of the mitochondria come from the mother. Each individual one is a copy of the mitochondria found in the egg carried by the birth mother. It seems probable that, even in the unfertilized egg, mitochondrial aging is taking place. There is some evidence for this based on the observation that the offspring of older female fruit flies, houseflies and mice all have a shorter life span than those from younger mothers. The age of the fathers shows no effect.

As fruit flies age they lose the ability to fly and begin to have difficulty even walking around. As this happens there is a large loss of the genetic material in the mitochondria, the mitochondrial DNA. When this loss occurs, there is no way the mitochondria can continue to carry out their very important function of energy production. Their information source, their library if you will, is lost. It is like a computer that has lost all of its programming with no back up.

We know that there are DNA repair enzymes in the nucleus. No one has ever found DNA repair enzymes of any kind in the mitochondria. There is not very much DNA in a mitochondrion, a lot less than what you find even in bacteria, but it is absolutely critical for maintaining the function of this organelle. This is a real puzzle since even simple bacteria have DNA repair enzymes. It makes no sense to me, unless possibly mitochondrial DNA loss is the predetermined biological clock for aging. Denham Harman first proposed the idea that the breakdown of mitochondria represents the biological clock for aging in 1972 (Journal of the American Geriatrics Society, vol. 20, pp. 145-147). There must be a reason for this very unusual phenomenon of mitochondrial decay, but we do not know for sure what it is. There is some evidence for why this occurs in Chapter 12.

This loss of mitochondrial DNA occurs much earlier in the Swedish C strain of Drosophila that lives only 30 days at room temperature than it does in the Oregon R strain which lives 50 days at the same temperature. In both the Swedish C and Oregon R strains, the rate of mitochondrial DNA loss depends upon the environmental temperature. The rate of loss is much faster at higher temperatures than at lower temperatures. It sounds as if we are back to the "rate of living" theory of aging. It seems that the mitochondria become damaged with time based on the rate of living.

Mitochondria produce almost all the energy used by a fruit fly for movement and basic metabolism. In mammals and birds the heat produced by mitochondria maintains the body temperature. Many older people feel cold much of the time. A health care worker has told me that one of the most welcome gifts for a nursing home resident is a sweater. It seems that when the mitochondria wear out, the organism or individual wears out. The real source of the problem appears to be the loss of the mitochondrial genetic material. Since there are no repair mechanisms for this loss, the only viable approach seems to be the prevention of this damage. To address this scientists have tried to come up with possible causes of the damage.

Chapter 5

THE FREE RADICAL THEORY OF AGING

In 1956, a former industrial chemist, scientist and physician named Denham Harman published the free radical theory of aging. He has told me that he almost gave up on the idea of publishing it but went ahead with it anyway. Quoting from the article says it best: "Aging and the degenerative diseases associated with it are attributed basically to the deleterious side attacks of free radicals on cell constituents and on the connective tissues." This was a monumental statement which launched a whole wave of studies into, not only the aging process, but all of the diseases associated with aging and other diseases too.

Free radicals have been implicated in just about every noninfectious disease one can think of. The list includes heart disease, stroke and cancer. These are the three major killers. Inflammation and pain are also believed to be due to free radical mechanisms. Thousands of scientific and many medical articles have been published concerning the role of free radicals in biology and disease since Dr. Harman, whom I know as a very modest man, first proposed this concept. He has been nominated for the Nobel Prize for this work, but he has not, unfortunately, received it. The point is that he started a revolution in how people think about what is going on in both aging and disease states.

Definition of a free radical

To understand what a free radical is one has to know that all things in the universe are made up of atoms. Atoms have a center, the nucleus. Spinning around the nucleus are electrons. This is much like the planets spinning around the sun. The smallest atom is hydrogen and one of the largest is lead; uranium is even larger. That is why one would want to make bullets out of lead rather than hydrogen. The military people have heard about uranium atoms being even larger than lead and that is why they have fairly recently made armor penetrating projectiles out of these atoms.

Every atom has its electrons and they normally exist in pairs. These electrons can pair up with other electrons from other atoms to form molecules. This is their happy state, as they say. When they are unpaired, there is trouble. Water, H_2O, is two atoms of hydrogen paired with one atom of oxygen, making it one of the smallest molecules. Millions of atoms can pair up to form gigantic molecules such as DNA, collagen proteins in your joints, and even man-made plastics.

Collagen is a good example since it is the most numerous of all the proteins in the body. It is everywhere: in the bones, the joints, the skin, muscles and all the organs. It is like glue holding everything together. The tendons are mostly collagen in the form of strong fibers that allow you to move about and even run marathons. Collagen, like all proteins including enzymes, is made up of amino acids which, in turn, are made up of atoms of hydrogen, carbon, nitrogen and oxygen.

The code for making amino acids and their larger molecular products, proteins, resides in the DNA molecules of the cell's nucleus. Collagen in the body is seldom replaced. One theory of aging is that changes in collagen are a problem. This has been called the collagen theory of aging. Collagen becomes old and brittle with time and this is why there are all sorts of joint and flexibility problems with increasing age. Broken fingernails are probably due to the aging of collagen. All of this may be true and can certainly be the cause of serious disability problems, but it seldom leads to death. Is it the basic cause of aging? Perhaps.

Thousands of different molecules exist in all living things. When any one of these molecules has an unpaired electron, it is called a free radical. In order for electrons to have a normal existence they must be in pairs. If not, they will seek out another electron no matter what the consequences. They will combine with any electron they can find. A free radical is any molecule that has an unpaired electron and is looking for trouble. One could think of it as a molecule with a bad attitude.

Damage is done when these free radicals react with other molecules with which they should normally have nothing to do. It can be thought of as assault at the molecular level.

Most free radicals are believed to be produced when the cell oxidizes or burns sugar or fat molecules in order to produce energy. These are called oxidative reactions simply because they involve the consumption of oxygen. The process is a chain reaction and is, therefore, referred to as the respiratory chain. Free radicals are produced during the utilization of oxygen in the respiratory chain. This takes place in the mitochondria of every cell in every organ. This is normal and not harmful as long as these free radicals stay in the chain. It is the ones that escape the chain that cause random damaging reactions. Denham Harman has called these the "free" free radicals.

One could think of all of this as an old time prison chain gang with men chained together while on a work detail. Imagine one of them with a really bad criminal background escaping. He would be the "free" free radical and might do all kinds of horrible things before being recaptured. In short, it is not the free radicals in the respiratory chain that are the problem; it is only those that escape.

Over 90% of the oxygen used in the cell is processed in the mitochondria since these are the power plants of the cell. If free radicals cause damage to other molecules then this is where one would expect to find most of that damage. With aging there is considerable damage to the mitochondria. Especially noticeable in fruit flies is the loss of their genetic material, the mitochondrial DNA. This loss could be the result of free radical damage and would be permanent since there are no DNA repair enzymes in the mitochondria.

Oxygen and free radicals
If free radicals are the reason for aging and since most free radicals are derived from oxygen, then the amount of oxygen present in biological tissues should affect the rate of aging. The dangerous free radicals are almost all derived from molecular oxygen. It is estimated that about 10% of all the oxygen taken in through respiration ends up as free radicals. A commonly used strain of mice, CFW female mice, normally live an average of 500 days in ordinary air, which is 21% oxygen. When they were given air containing 70% oxygen, they lived an average of only 200 days. When their surroundings were increased to 100% oxygen, they lived an average of only 111 hours. Yes,

hours not days. When the pressure of the oxygen was increased by placing them in steel chambers their survival was even worse. At an oxygen pressure of ten times normal atmospheric pressure they lived for only 18 minutes. This is a shocking result and indicates that oxygen, which is necessary for life, can be extremely toxic.

This study of oxygen toxicity was sponsored by the United States Air Force. It was published in the American Journal of Physiology, vol. 192(3), pp. 563-571, in 1958. Pressurized oxygen, rather than air, would save a lot of weight in a spacecraft or an aircraft flying at high altitudes, but the results would be disastrous. The results of this research also offer strong evidence for the involvement of oxidative molecules in the aging process. Almost all of these reactions are believed to be caused by free radicals. The results of this study are almost incredible since one usually does not think of oxygen as being harmful.

Could there be too much oxygen in ordinary air today? Did the Biblical patriarch, Methuselah, in the first book of the Old Testament, Genesis, really live to be 969 years old? Was there less oxygen in the air in those days? The paleontologists tell us that during the Jurassic period of time before the Cretaceous era that the oxygen level in the air was 12 percent. About 100 million years ago, in the middle of the Cretaceous age, the atmospheric oxygen levels started rising, until the present concentration of 21% was reached 50 million years ago. Oxygen levels in the atmosphere changed considerably during these time periods. Perhaps there have been other changes in the oxygen content of the atmosphere during more recent periods of geologic time that have affected the basic aging process. Unfortunately, no one has tested this simple idea.

Radiation produces free radicals

There is other evidence supporting the free radical theory of aging. Without a doubt the most striking example of free radical damage was the nuclear bomb drops on Hiroshima and Nagasaki, Japan during the Second World War in 1945. The explosive force of these bombs was enormous, but expected, based on a test in the New Mexico desert. However, the accompanying radiation damage to humans was totally unexpected and devastating. No one had predicted that such a quantity of radiation would be released, and all were horrified at the consequences.

The suffering of the survivors of the blasts was enormous, and no one had the knowledge to treat it. The Atomic Energy Commission (which no longer exists) went into high gear to try to understand what had happened. It took time, but the final conclusion was that it was a massive wave of free radicals produced by the radiation from the bombs that had caused the problems. These "free" free radicals made many people very ill and killed even more.

It has since been shown that high-energy radiation such as gamma rays and x-rays can produce very large numbers of free radicals in biological organisms. The exact number is known. Two trillion free radicals are produced in only one gram of water by only one rad (a unit of radiation exposure) of radiation. It takes a single dose of 500 rads of radiation to produce accelerated aging in mice, which reduces their life span by half, and makes them look very old at a young age. This amount of radiation does not kill them, but it cuts their life span by a considerable amount, 50%.

Scientists from the Atomic Energy Commission and elsewhere were able to show that even comparatively small doses of low energy radiation, such as x-rays, could turn young mice into old looking mice and also greatly reduce their life span. Attempts were made to find chemical agents to protect against this process, but nothing effective was ever found.

Most humans are exposed to only 0.2 rads of environmental, or background radiation, per year. That would mean that background radiation produces 400,000,000,000 free radicals per year in each gram of your tissues. This is a large number, but it is probably small compared to the number of damaging free radicals produced during the normal metabolism of oxygen in the body. With all of this damage going on every day, it is truly amazing that we live as long as we do.

Is there anything a person can do about background radiation? There is not very much. The geologists tell us that at the center of the earth is a gigantic nuclear reactor based on the decay of uranium. It is really hot down there. This is why there are volcanoes made up of molten rock and metals from the center of the earth, and why in the deep diamond mines in South Africa people need air conditioning in order to survive.

One of the products of this giant nuclear reactor is radioactive radon gas. Rocks and bricks release radioactive radon gas, so it is better to have a wooden house. It is also released from the earth into your basement and house. A test kit for radon gas can be purchased for about $5 if there is a concern about radon in the house or basement. Painting the basement with latex paint usually does away with the problem. If not, then it is necessary to put in a ventilation system to get rid of it. The experts claim that high concentrations of radon gas can increase the incidence of lung cancer.

The fact is that all of the food, water, and air that humans breathe is slightly radioactive. This is generally referred to as background radiation, but some people are exposed to more of this radiation than others.

About half of the background radiation comes from outer space. In Denver at 5280 feet of altitude the exposure is about twice as much to cosmic radiation as it is in New Orleans which is at an altitude close to sea level. The curious thing is that people who live at high altitudes tend to have less cancer than those who live at lower altitudes. This has led some scientists to make the astounding suggestion that a little radiation is probably good for you.

Surprisingly, low levels of radiation do slightly prolong the life span of mice when compared to mice that receive no radiation exposure at all. The same thing is true for fruit flies. This does not necessarily mean that the free radical theory of aging is wrong. This apparent contradiction, however, must be explained (Chapter 19). We do not know how many free radicals are produced per gram of tissue by the metabolism of oxygen in the mitochondria of cells. There is a way, however, to make a pretty good guess. Instead of 500 rads of radiation to reduce the average life span by more than half for a mouse, it takes 50,000 rads of radiation to reduce the average life span of a fruit fly by more than half. This is a difference of one hundred fold. Why the big difference? Consider that the basic metabolic rate for a fruit fly is much higher than it is for a mouse. If most of the free radicals are produced by ordinary metabolism, then a fruit fly will have a lot more free radicals per gram of tissue than a mouse. Therefore, radiation induced free radical production would have to be much higher in a fruit fly in order to make a difference in its life span. A dose of 500 rads would hardly make a difference to a fly, whereas 50,000 rads would add to the already enormous burden of free radical production that they experience.

Since we know the number of free radicals produced by a given amount of radiation, thanks to the Atomic Energy Commission, one can guess what the normal amount of free radical production must be in a fly. That would be 50,000 times the number of free radicals produced by 1 rad of radiation which is two thousand billion radicals per gram of tissue. That means that fruit flies have 50,000 times 2,000 billion, or 100,000 trillion, free radicals per gram of tissue at any given time. It must be kept in mind that this is only an educated guess.

These are, of course, huge numbers, and one can start to appreciate why a fruit fly lives such a short time. They are being bombarded with free radicals. For a mouse this number of free radicals would be a hundred times less, based on what we know about their radiation sensitivity. The CFW mice live 500 days. This is a little more than one hundred times the life span of fruit flies kept at about the same body temperature as a mouse. It does seem to suggest that these two totally different organisms have differences in radiation sensitivity proportionate to almost the same differences in their life span. In short, it takes 100 times more radiation to reduce the life span of a fruit fly than it takes to do the same thing to a mouse. It seems that the reason for this is due to the differences in the basic rate of free radical production in these two different organisms.

It is also interesting to note that up until 1945, radiation in the form of x-rays was used successfully to treat various disease states, including cancer and even acne.

I remember a very nice man from New York City who told me that, while serving in the U.S. Navy in the Pacific during World War II, he developed a fungal infection around the base of his officer's cap. The navy physicians cured it by exposing his head to high doses of x-rays. It worked, but I am sorry to say that he eventually died of brain cancer.

High doses of radiation are still used today to treat various diseases, including cancer. In cancer treatment, it is the high levels of free radicals created by radiation exposure that kill the cancer cells. Fortunately, this exposure is highly controlled and usually localized so that the rest of the body is only moderately affected.

Light produces free radicals

Exposure to light also produces free radicals. Ultraviolet light exposure is the most effective way of doing this. This is the high energy light from the sun that gives us sunburns and skin cancer. Chemists have known about this for a long time and have used both ultraviolet and visible light to produce free radicals in all kinds of chemical reactions.

Included in these reactions is the making of porcelain crowns for teeth. I just had this done. My dentist applied a layer of liquid to the tooth and then hardened it with a small ultraviolet light source. The free radicals produced by the ultraviolet light started a chemical reaction that turned the liquid into a solid. By applying additional layers he was able to produce what looks like a perfectly normal tooth. Incidentally, the cost was a lot less than a regular crown. It is possible, however, that this procedure caused accelerated aging on the inside of my mouth.

The conclusion here is that free radicals can greatly increase the rate of aging if they come from outside sources. However, internally produced free radicals are probably also a major, if not the major, source of biological aging. What can be done about this? This is the subject of the next few chapters.

Chapter 6

METAL IONS AND TWO PROTECTIVE ENZYMES

Metal ions also affect the rate of free radical production. All of us are familiar with metals such as iron, copper, aluminum, lead, mercury, zinc, cadmium, silver and gold. Less familiar metals include sodium, potassium, cobalt, calcium, manganese and magnesium. This latter group is not as familiar to most people as metals because they are usually seen only as ions. Ions of metals are atoms that have an electronic charge and tend to combine with other atoms of an opposite electronic charge. The atomic ion of the metal sodium, for example, is usually found in nature to be combined with the atomic ion of chlorine to form sodium chloride, common table salt. The reason for this is that most metals are very unstable and reactive. As a result of this, metals are usually found in nature in their ionic form. They readily convert to their ionic form if found in a pure metallic state.

Gold and platinum are notable exceptions since they are quite stable in the pure metallic state. Silver is fairly stable, but it does oxidize, as every kid who has had to clean the family silverware knows. Iron easily goes to the ionic form that can be recognized as rust on cars and other products.

I remember the greatly embarrassed chemistry professor who was showing his class metallic sodium. Stored in a jar filled with kerosene, it is stable, but when exposed to air, it begins to fume and will react with just about anything with which it comes in contact. He dropped it. The sodium metal pellet proceeded to eat a hole in the demonstration table, went through 3 shelves, the floor and was on its way to the basement floor below before it finally dissipated. This is one example of why you do not see most metals in their metallic form but only as salts made up of the ions of these metals.

Anyone who owns a car knows that iron is very reactive. Coats of paint help to delay the process, but eventually rust takes over and ruins the car. The point is that almost all of the metals on the earth's surface exist in ionic form. Only man and volcanoes have been able to convert them to their metallic form as seen in the familiar metal products of iron, aluminum, copper and silver.

Metal ions

In biological systems all metals exist in the form of ions. They have an electronic charge and can, and do, participate in all kinds of chemical reactions. It should be emphasized that while metal ions have an electronic charge, they are not free radicals. All of their electrons remain in pairs with no free electrons available for easy chemical reactions, as is the case for free radicals. They can be independent and unattached, and this is when they create problems, but in living organisms they are almost always attached to proteins.

One of the most familiar metal proteins is hemoglobin, the protein inside the red blood cells that carries oxygen throughout the body. It is red in color which is why blood is red. It is able to carry oxygen only because it contains ionic iron. Other less familiar proteins that contain metal ions include most of the enzymes. These are the proteins that carry out all the important biological processes of energy production, growth, repair, and the killing of viruses, bacteria, fungus and even cancer cells. Even the vitamin B12 contains a metal ion, cobalt.

The metal ions are there in proteins because the chemists claim that they lower the activation energy for all kinds of reactions. That means that it takes a lot less energy to do a job if there is a metal ion working. It is like waking up on a January morning with three feet of snow in your driveway. One could use a shovel, but a seven horsepower snow blower would be better. The snow blower is analogous to the metal ion attached to a protein to form an enzyme. The lowering-the-activation-energy concept is equivalent to using the snow blower rather than the shovel. Without metal ions, life as we know it, would not be possible.

The problem is that these same metal ions, which can be so helpful, can cause an incredible amount of damage if they are free, and in the wrong place, and not attached to a protein enzyme.

Unattached, or free, metal ions enormously increase the rate of production of free radicals. Again, chemists have known this for years. In order to get a free radical reaction to go a lot faster, one can simply sprinkle in a few metal ions. The chemists call these catalysts. Metal ions, acting as catalysts, affect the aging process in both a positive and a negative way.

Antiradical proteins
The positive way for metal ions to influence the aging

process is through the action of protective antiradical proteins. All of these proteins contain metal ions. These are generally referred to as antioxidant enzymes. Most of the free radicals in living organisms are believed to be oxidizing molecules primarily comprised of oxygen in one form or another. The antioxidant enzymes are all basically designed to prevent the formation of the worst free radical of them all, the hydroxyl radical. All of the protective enzymes have metal ions at their reactive centers. Without these protective enzymes, human beings would probably never live to see their first birthday.

One of these protective antioxidant enzymes is catalase. It decomposes hydrogen peroxide into water and oxygen. Hydrogen peroxide is a strong oxidizing agent, which means that it contains oxygen and can react with just about anything. It can be purchased in any drug store and be used as a disinfectant and cleaning compound. A cut can be disinfected with hydrogen peroxide. The bubbles coming off are oxygen produced by the catalase molecules in the blood.

During normal metabolism in humans, and all other organisms, a great deal of hydrogen peroxide is produced. It is estimated that the average human produces over a ton of it during their lifetime. A ton of hydrogen peroxide would, of course, be fatal to any living organism if it were not removed in a nontoxic manner. Hydrogen peroxide is a strong oxidizing agent. Basically this means that it can combine with almost any other molecule. Hydrogen peroxide is so powerful an oxidizing agent that it has been used as a rocket fuel. The enzyme, catalase, removes it from the body by producing oxygen and water. At the center of the catalase molecule is an iron ion. Without it the enzyme would be unable to function. This is one example of the importance of metal ions in providing protection against powerful oxidizing agents.

If hydrogen peroxide is left to itself, it does not produce free radicals. In the presence of certain free metal ions, such as iron or copper, it can break down into two hydroxyl radicals. Each of these free radicals is composed of one atom of oxygen and one atom of hydrogen with an unpaired electron just looking for trouble. There is nothing more reactive than an unpaired electron looking for a mate. In addition, the hydroxyl radical is the worst of them all. It is highly reactive, and because of its small size, it can go almost anywhere.

What makes this chemical process even worse is the involvement of these free metal ions in biological tissues. It should be noted that a free metal ion is not a free radical. Ions and radicals are two different things. If there are any unbound or free metal ions near the hydrogen peroxide molecules present in the cell, these ions will cause the hydrogen peroxide molecules to break down into free hydroxyl radicals each carrying that dangerous unpaired electron. Iron, copper, manganese and titanium ions will do this. So while the enzyme catalase with an iron ion located at the center of its reaction site can break down hydrogen peroxide into the harmless products of oxygen and water, metal ions by themselves can do the opposite. They can cause the release of very toxic and damaging free radicals in the form of the hydroxyl and other free radicals. Research has clearly shown that DNA molecules are broken, or even totally destroyed, by hydroxyl radicals. This is especially dangerous to the mitochondria since they have no DNA repair system.

In a sense this is a paradox. Metal ions attached to an enzyme can be life saving, but by themselves, they can cause problems. Copper and manganese are two other metal ions that can accelerate the production of free radicals, and yet they are at the active site of one of the most important antioxidant enzymes, superoxide dismutase. Superoxide dismutase catalyzes, or stimulates, the decomposition of superoxide free radicals. The superoxide free radical is composed of two atoms of oxygen with one unpaired electron. It is the ordinary oxygen molecule that one inhales with one difference; it has an extra electron attached to it. This is why it is called a free radical.

There are two forms of the enzyme superoxide dismutase. One contains a copper ion at its reactive center and is found in the cytoplasm (the cellular soup, if you will) of the cell. The other has a manganese ion as its active site and is found only in the mitochondria of the cell. These two forms of superoxide dismutase (abbreviated as SOD) are found in all the cells of all the organs of the body of all organisms that have been examined. This is in contrast to what is known about the distribution of the enzyme catalase. Catalase activity levels are high in liver and kidney cells, but other organs, such as the muscles and brain, have very little activity. This suggests that SOD is a very necessary antioxidant enzyme in all cells while catalase does not seem to be as universally required, at least at high concentrations.

The really important thing about superoxide dismutase is that a separate and unique form of it, the manganese SOD, is present in all mitochondria. It must be there for a reason. Since mitochondria are the site for the oxidative chain where 90% of the body's oxygen is consumed, it makes sense that any escaping superoxide radicals would have to be removed as soon as possible before they attack important molecules such as the DNA in the mitochondria.

Superoxide radicals can combine with hydrogen peroxide to produce those very reactive hydroxyl free radicals. In order to avoid this, the cell uses both the enzymes, SOD and catalase, to remove these oxidative molecules.

Changes with aging in antioxidant enzymes

The activity levels of catalase and SOD decline with age. Total catalase drops by more than 50% in fruit flies with increasing age. A similar decline is observed with age in the blood of CFN male rats with most of the decline occurring before 300 days of age when the rats are young adults. The ability to make new catalase molecules is greatly reduced in old C57BL/6J mice.

A decline with age in the enzyme, superoxide dismutase, in the cytoplasm has been reported for mouse and rat liver and for mouse brain. Other investigators have reported no change with age in SOD in the cytoplasm for human erythrocytes (red blood cells), fruit flies and for various rat and mouse organs. In contrast, mitochondrial SOD has been reported to decline with age by 33% in rat liver and by 21% in the fruit fly.

There is no difference in SOD activity levels in twelve different primate species of greatly different life spans. This is a problem when considering SOD to be of significance in the aging process. However, the ratio of SOD activity to specific metabolic rate (rate of energy production) increases with increasing life span in these different species. No significant differences were found in either mitochondrial or total SOD activity levels in two different wild-type strains (Oregon R and Swedish C) of Drosophila whose life spans differed by 40%. The mitochondrial SOD activity, however, did decline with age in both of these strains, but not greatly.

In conclusion, the total amount of SOD activity does not appear to be related to life span in either animals or the fruit fly. The decline with age of the SOD in mitochondria, however, fits in with the idea that mitochondria may represent the site for biological aging. Whether or not the age related changes in SOD or catalase are of a sufficient magnitude to contribute greatly to the basic aging process remains unknown. It is, nevertheless, clear that these two protective enzymes are necessary for the defense of biological organisms against free radicals.

Chapter 7

LIPID PEROXIDATION, A THEORY OF AGING

Lipids are fats. There are two kinds of fats, saturated and unsaturated. Butter, coconut oil and lard are examples of fats that are primarily saturated. Fish oil, corn oil, cottonseed oil and just about any kind of vegetable oil are examples of unsaturated fats. A saturated fat has only one set of electrons between its carbon atoms. This is called a single bond. Unsaturated fats have two or more sets of electrons between their carbon atoms. These are called double or triple bonds. You would think that a double bond would be stronger than a single bond, but this is not the case. All of those extra electrons in a double or triple bond are looking for other electrons with which to pair. The single bond electrons of saturated fat are quite happy with each other and have little interest in other electrons.

The breakdown of lipids or fats occurs when the double and triple bonds are exposed to oxygen. This is called lipid peroxidation simply because, in the process of lipid oxidation, peroxides are formed. These peroxides are similar to the hydrogen peroxide found in any drug store except they are larger and more complicated. In general, most oxidation reactions (this is basically biochemical rust) involve the addition of oxygen to another molecule. This is not always true, but for the sake of this discussion, one can assume it to be so. Lipid peroxides can be very large molecules, but they all have those two atoms of oxygen just as hydrogen peroxide does. When fats go rancid lots of lipid peroxides are produced. These peroxides are the reason for the very unpleasant taste of rancid fats and oils, especially noticeable in seafood that is not fresh. These are all oxidation products. One could think of them as food rust products similar to the rusting of a car by the formation of iron oxides due to oxidation by atmospheric oxygen.

Aging and rancid fats
The lipid peroxidation theory of aging says, basically, that biological organisms go rancid with time, and this is the reason for aging. There is a lot of suggestive evidence for this idea, and it may or may not be correct.

Many books and scientific papers have been written on the subject of lipid peroxidation. The main reason for this is that edible oils and fats are an article of commerce of enormous importance worldwide. The processing and, more importantly, the preservation of these products are of great economic concern. The food chemists have, therefore, produced the greatest volume of research in this area. They report, for example, that oxygen is seven times more soluble in oil than it is in water, and that one of the most important things to do to keep oil from going rancid is to remove oxygen from the product. Another very important thing to do is to keep out any kind of metal ion, especially iron and copper. All of this sounds familiar. It gets back to two of the major factors previously discussed above involving free radical damage in biological systems. Free radical damage is the principal, if not the only, reason for loss of quality in all food sources containing fats and oils.

I remember the comedian, George Burns, when he was quite old being asked if he ate natural foods. His reply was "No", and said that, at his age, he needed all the preservatives that he could get. The food scientists know all about preservatives and have very clever ways of controlling free radical damage to foods by blocking oxygen and free metal ion damage. Still, it is clear that fresh foods taste better.

The lipid peroxidation theory of aging says that we all have a problem with increasing age due to the oxidation of polyunsaturated lipids in our cells. Most of these lipids are located in the cellular membranes, including the membranes surrounding the mitochondria. The oxidation reactions occurring in the mitochondria take place within these membranes. It, therefore, would be expected that at least some of the polyunsaturated lipids in the membranes would be accidentally damaged during normal oxidative metabolism.

This damage is the result of the reaction of oxygen with the fats and oils in the membranes. It is a process that produces free radicals and lipid peroxides, both of which then contribute to a destructive chain reaction. This process proceeds with great speed. Metal ions such as iron and copper greatly accelerate this already rapid process. The superoxide and hydroxyl free radicals are involved in the process, as one might expect. Even smog pollutants, such as ozone and nitrogen dioxide, have been shown to be involved in lipid peroxidation, and this is probably the reason that they are so harmful. Certain commonly used chemicals, such as carbon tetrachloride (used for dry cleaning and as an organic solvent), can also be involved in the oxidative breakdown of biological lipids. The chemical literature on the subject of lipid peroxidation is extensive and quite impressive.

As far as the involvement of lipid peroxidation in the biological aging process is concerned, surprisingly few measurements of the products of peroxidation have been reported. In addition, not all of these reports have been consistent. One of the products of lipid peroxidation is the chemical malonyldialdehyde. This has been found to increase with age in Wistar rats. On the other hand, there is a considerable decrease with age in the concentration of malonyldialdehyde in male Sprague-Dawley rats. An increase in the concentration of lipid peroxides has been reported for fruit flies.

In the most comprehensive study on this subject, however, lipid peroxidation did not increase with age in C57BL/6J mice, even in very old mice. In this report none of the organs examined, including the heart, liver, kidney, and brain, showed any significant change with aging. From this point of view, the scientific evidence shows little support for the lipid peroxidation theory of aging. This is of concern since lipids are such a sensitive site for free radical and oxidative damage. It throws the whole free radical theory of aging into question.

One possible explanation for this lack of support for the lipid peroxidation theory of aging is that our analytical techniques are just not good enough to show the involvement of lipid peroxides in the aging process. This seems unlikely. If these techniques work for vegetable oils and other foods, why would they not work for biological systems? Another possibility is that these reactions are so fast that we cannot catch them, so to speak. A reason for this is presented a little later on.

Another reason for the lack of evidence could be that

the membrane defenses and repair mechanisms are so overwhelming that lipid peroxidation is not a major factor in the aging process. The evidence for this is good and is presented in the following sections.

Vitamin E and glutathione peroxidase
The membranes of the cell, mitochondria, and other cellular organelles are constantly repaired and replaced. In addition, there are some powerful protective systems in place to protect the lipids in membranes. One of these is the enzyme glutathione peroxidase, and another is vitamin E. Most of us are familiar with vitamin E. It is added to many breakfast cereals and other foods. High concentrations of it occur naturally in all vegetable oils. It is an article of commerce and can be purchased in any drug store. Two of the problems associated with vitamin E difficency in animals are sterility and impotence. Many people, therefore, believe that taking supplements of this vitamin will avoid these problems and even improve their sex lives. The fact is that, unless a person eats a totally fat free diet, vitamin E defiency is very unlikely.

Vitamin E is the king of all the molecules labeled as antioxidants. It reacts with both oxygen in all its forms and with free radicals. It prevents edible oils from going rancid. It is one of the few fat-soluble vitamins that are not toxic when taken at high doses. Vitamins A and D, for example, are two fat-soluble vitamins that can be very harmful if taken in excess. This is not true for vitamin E, as far as I know. Still, one should think of this as a possibility. I have found that very high doses of vitamin E do reduce the life span of Drosophila. However, this is unpublished data.

To a biochemist interested in aging, vitamin E sounds like a miracle compound. What more could you ask for?

It is fat soluble, a powerful antioxidant, reacts with free radicals, prevents lipid peroxidation, a natural molecule and apparently nontoxic. It has been fed to fruit flies and rodents for their entire life span to see what would happen. The results were that nothing changed at all with aging in rats. Elevated doses of vitamin E in the diet do not significantly prolong the life span of fruit flies or rats or any other organism that I am aware of.

There is, however, one claim from a very reputable Drosophila researcher of a 15% increase in the life span of fruit flies when vitamin E was given at a high dose. I found no such increase even when I combined vitamin E with dietary selenium which is another very important element involved in protection against lipid peroxidation. As one can imagine, the results of these studies were a disappointment to the people involved. Nevertheless, they greatly improved our knowledge of the basic aging process. The conclusion here is that small amounts of vitamin E in the diet are sufficient to provide protection against lipid peroxidation.

Another important antioxidant enzyme is glutathione peroxidase. This is one of the most unusual enzymes in the cell. It has a rare element, selenium, which is a semimetal ion, at its center of action. Selenium is called a semimetal rather than a metal because it can be present in molecules either as a metal or as a nonmetal. As far as I know, glutathione peroxidase is the only enzyme to contain selenium. Selenium is an essential nutrient, and a rich source of it in the diet is Brazil nuts. It should be mentioned that excess selenium in the diet can be quite toxic. The reason for this is not known. Occupational exposure has caused pallor, nervousness, depression, and garlic odor of the breath and sweat.

Additionally, glutathione peroxidase requires a companion or cofactor to work: glutathione. Glutathione is a tripeptide, which means that it is composed of three amino acids, one of which is cysteine. Cysteine is also unusual in that it contains a sulfhydral group that is composed of one sulfur and one hydrogen atom. Sulfhydral groups are well known to chemists to react readily with free radicals, peroxides and other oxygen containing molecules.

A large decline in glutathione peroxidase would probably have very serious consequences for any organism for both disease states and the basic aging process. It is of interest that one of the symptoms of selenium deficiency is muscular dystrophy. Since selenium is at the reaction site of glutathione peroxidase, lack of dietary selenium for the synthesis of glutathione peroxidase is most likely the basis for this muscle wasting condition.

As can be imagined, scientists interested in human muscular dystrophy have given this subject a significant amount of attention. The surprising result, however, is that in mice with genetic muscular dystrophy the levels of glutathione peroxidase are the same as in normal mice in almost all organs including liver and heart. In muscle where you would expect to see a decline, the activity levels are actually higher than in normal mice. During the basic aging process in animals and in humans there is a normal loss of muscle protein with age, even for a determined body builder. There is, however, no real decrease in the number of muscle cells. Decreased muscle mass in older people is, therefore, probably not due to selenium deficiency, but rather to the basic aging process and, perhaps, a lack of exercise.

Very little work has been published on the changes in glutathione peroxidase with aging. There are other peroxidases in biological organisms and it is difficult to tell whether or not scientists should be looking at glutathione peroxidase or total peroxidase activity. One way to report results is to describe the changes in total peroxidase activity. It is a pretty good guess that some, or even most, of this activity is due to glutathione peroxidase.

When the changes with aging in total peroxidase activity were measured for fruit flies the results were shocking. The values increased with age in young adults, but dropped like a rock as the flies grew older for both male and female flies with male flies showing the greatest decline. The value for old male flies was about 20% of that of young adult males; the decline for females was about 50%. Unfortunately, we do not know how much of this decline was due to glutathione peroxidase activity. Along with these changes was a very large increase with aging in the number of peroxide molecules in the old fruit flies. The authors of this paper were Armstrong, Rienhart, Dixon and Reigh, published in AGE, vol. 1, pp. 8-12, 1978. It must be kept in mind that they did not measure glutathione peroxidase alone but all peroxidase activity.

The published scientific literature for changes in peroxidase with aging in animals is sparse. One study found a marked increase in the rate of malonyldialdehyde production (which indicates a lack of peroxidase activity) with age in mitochondria, whereas another study reported no changes in lipid peroxidation levels in freshly isolated mitochondria from rats of various ages. In view of the potentially great importance of peroxidase, and glutathione peroxidase in particular, in disease states and human aging, it is disturbing that so little research has been done in this area. The results that we do have are conflicting and inconsistent.

Perhaps the reason for this is that increasing the dietary intake of one or all of the factors involved in the glutathione peroxidase system does little or nothing towards reducing the rate of aging in all of the biological systems used to study it. Increasing the dietary intake of selenium, glutathione, cysteine and other sulfhydryl compounds makes no difference in the overall life span when compared to a normal healthy diet. One would think that these compounds would at least improve the life span of fruit flies since they show the greatest changes with aging in peroxide levels. However, they do not. It is possible that the lipid peroxidation theory of aging is wrong. There are great inconsistencies in some of the data and a general sense that something is missing. The fact that extra vitamin E in both fruit flies and rats does little or nothing to increase life span is also very puzzling. It could be that the available knowledge is simply inadequate for understanding the process. More evidence may be necessary.

Age Pigments

There is some additional evidence. As early as 1886, it was reported that age pigments appear in the cells of older organisms, including man. These can be seen with a microscope and were given the horrible sounding name of lipofuscin. Your pronunciation of this word is as good as mine. They have also been called lipopigments, age pigments, dense bodies, fluorescent granules, fluorescent chromophores, fluorescent bodies, fluorescent products and even some have named them cellular garbage. They appear to be dark, dense and large particles under the microscope. Under ultraviolet light (which humans cannot see), they give off visible light that can clearly be seen. They literally glow in the dark. This emission of light of another kind or wavelength is called fluorescence. Hence, the name, fluorescent granule or fluorescent products as some prefer to say. The term, cellular garbage, reflects the fact that these objects have no known useful function.

In addition to lipofuscin, there is another age pigment, ceroid. Under the microcope it looks almost identical to lipofuscin, and it also is fluorescent. Ceroid is seen in some disease states whereas lipofuscin seems to be entirely related to the aging process. Biochemicaly, they both are composed of fats, proteins and metal ions. Chelating agents can dissolve ceroid but not lipofuscin. Chelators are compounds that surround metal ions and tightly bind to them. "Chela" is the Greek word for claw. Ceroid has a high concentration of iron and calcium ions, whereas lipofuscin has a high content of zinc. It has been suggested that ceroid is just an early form of lipofuscin. They both seem to be forms of cellular garbage and both have a high content of oxidized fats (fats are usually referred to as lipids). Since some of the oxidation products of pure lipids are fluorescent, this is believed to be the reason that both ceroid and lipofuscin are fluorescent.

Both ceroid and lipofuscin are generally believed to orginate from organelles within the cytoplasm of the cell. One of these organelles is a mitochondrion; lysosomes are a second group and microsomes a third. Lysosomes are small organelles that contain a number of digestive enzymes used to remove debris from the cell. Microsomes are organelles which process and remove harmful chemicals that the cell might be exposed to. Lysosomes and microsomes do not contain DNA as the mitochondria do. Membranes containing unsaturated lipids surround all three. These membranes are believed to be the sites where lipid peroxidation occurs. It is believed that lysosomes engulf the damaged membranes of other organelles and also engulf damaged lysosomes in an effort to remove them. This effort sometimes fails. What remains behind is either lipofuscin or ceroid. The reason for believing this is that both purified lipofuscin and ceroid always have some residual lysosomal enzymes present.

A great deal of evidence points to the mitochondria as being the source of lipofuscin and ceroid. Presumably more damaged or broken down mitochondria accumulate with age than damaged lysosomes or microsomes. There is not, however, universal agreement that this is so. There is, on the other hand, complete agreement that these two fluorescent particles are composed in large part of oxidized lipids and that they accumulate with aging.

The most complete studies have been done with the human heart. An older man or woman can usually have 6% of the total volume of all heart muscle cells filled with lipofuscin. This is an incredible amount and is a result of an accumulation rate of 0.6% for every ten years beginning at a young age. The same process occurs in human brain and testes but to a lesser extent. Similar results have been found for laboratory animals.

In fruit flies the lipopigment is not lipofuscin but ceroid. In the digestive cells of old flies an almost unbelievable 50% of the cellular volume is occupied by ceroid. Fruit flies have ceroid, but humans have lipofuscin. The suggestion has been made that the flies do not live long enough for the ceroid to be converted to lipofuscin. These age pigments are so similar that it is very difficult to tell them apart. Whether or not these differences are of any consequence to the basic aging process remains to be established.

The large amount of ceroid in fruit flies is consistent with the fact that we observe a large increase with age in the number of peroxides in fruit flies, whereas little or no increase is seen in these molecules with age in laboratory animals. There must be an enormous amount of lipid peroxides being produced in an aging fruit fly. This is consistent with the fact that they have a higher rate of oxygen consumption as compared to other larger organisms such as humans.

Even the WI-38 human fibroblast cells in culture bottles, discussed as a model aging system in Chapter 2, accumulate fluorescent pigments as they grow older. Over 40% of the older cells have ceroid or lipofuscin particles, whereas only a few percent of the young cells have them. The interesting thing about this study was that the cells with the greatest amount of fluorescent particles were unable to undergo further cellular division. This was true whether or not they were young or old cells. They just stopped growing and, in addition, were unable to use or incorporate a molecule necessary for DNA synthesis, thymidine.

It seems, therefore, that every system for studying aging, including man, shows a definite increase in the number of fluorescent pigments with age. The results with the WI-38 system suggest that this accumulation interferes with normal cell function. I do not know, however, if this is true for whole organisms. For example, there is no relationship between the large amount of lipofuscin found in human hearts from older individuals and heart failure. Those dying of heart failure have the same amount of lipofuscin as people of the same age who die from other causes.

Those children suffering from the human genetic aging disease of progeria (Chapter 3) have absolutely no lipofuscin in their hearts. None at all was found in the two cases examined at the ages of 7 years, 6 months and at 11 years of age. Considering that these individuals with progeria seldom live to be 20 years of age and have all the outward appearances of extreme old age, it is possible that the formation of ceroid or lipofuscin is really a cellular protective mechanism against the damaging effects of lipid peroxidation and aging. Perhaps, the genetic loss of this mechanism is the reason for the extremely rapid rate of aging seen in these children.

Another very puzzling piece of evidence concerns the rate of fluorescent product formation with aging in fruit flies maintained at different environmental temperatures. In this study (Sheldahl and Tappel, Exp. Geront. vol. 9, pp. 33-41, 1974), Drosophila maintained at 30 degrees Celsius had a lower rate for the formation of age pigments than flies maintained at 26.7 degrees Celsius. This was true for both male and female flies with female flies showing the greatest difference. If lipid peroxidation is a basic cause of aging, how could this be possible given that flies at 30 degrees Celsius age at almost twice the rate as those maintained at 26.7 degrees Celsius?

Perhaps flies at the higher temperature are so overloaded with peroxides that they are unable to control them as well as the flies do at the lower temperature. That is, of course, if one considers that fluorescent pigment formation might be a defense mechanism against aging. As far as I know, this idea has not been proposed before. These results bring into question the whole aging theory of lipid peroxidation and age pigments. The only thing I can think of which might support these results is that children with the aging disease progeria have no age pigments. One would have to conclude, therefore, that the age pigments of ceroid and lipofuscin represent a beneficial result on the part of organisms to defend themselves against lipid peroxidation. In short, they could represent evidence that the body is doing its work. Your explanation, however, is as good as mine, and that is why I have referenced this important paper.

In conclusion, age pigments occur in almost all living systems and greatly increase in concentration with time in the cells of organisms, except for those children with progeria. They represent the product of the lipid peroxidation process. It is generally believed that they represent cellular garbage and that they slow down or retard normal functions within the cell. This idea may be wrong. Whether or not age pigments or the process of lipid peroxidation which leads to their genesis are responsible for the basic process of biological aging remains unkown. Much more work needs to be done.

Chapter 8

ANTIOXIDANTS

An antioxidant is anything that reacts with oxygen. More exactly defined it is anything that reacts with oxygen-type molecules or oxygen-containing molecules such as lipid peroxides. The literature on this subject is enormous and cannot be completely covered in this book. Only a summary of the effects on the aging process is offered here, and a very large amount of information is, therefore, omitted.

Many antioxidants have been tested for their effect on the length of life. This is generally referred to as life span. These changes are reported in three different ways. Average life span is the average time that an organism lives. If, for example, one mouse lives to be 1,000 days of age and another lives to be 500 days of age, the average life span for the two mice would be 750 days. Median life span is the time where half the organisms are still alive and half are dead. In this simple example that would be 500 days. Maximum life span is the time when the last surviving organism in a given study dies. For our 2 mouse example that would be 1,000 days. As you can see, there can be large differences depending on how life span is reported.

Vitamin E

Vitamin E is a very powerful antioxidant, especially as far as protection of fats and oils against oxidation is concerned. Unfortunately, in spite of numerous claims, the evidence for any aging related benefit from extra dietary vitamin E is minimal. This is a good example for the suggestion that we should all be suspicious of claims of fantastic age related results. Years ago two respected scientists published the incredible claim that WI-38 human cells in culture could be grown to more than 110 divisions (when normally they divide only 50 times) if they were given high doses of vitamin E. This really seemed like the answer to the aging problem to many, including me. I had done these experiments years before but found no positive result. As a result I believed that I must have done something wrong. I ordered the same chemicals used by these investigators and repeated the experiment. After more than 6 months of research, I still had the same result: no change. No one else could repeat these amazing results. Finally, there was a confession of data faking. Without realizing it, some people still refer to this false experiment, since it is published in the literature. The fact remains that little or no improvement in life span in any model system has been achieved by high dietary supplements of vitamin E. There is, however, an enormous amount of literature supporting numerous claims of health benefits for increased dietary vitamin E. Just one of these is a report of greatly reduced cataracts in older people who take vitamin E.

Vitamin C

Perhaps the most widely known of all the antioxidants is vitamin C, also known as ascorbic acid. Unlike vitamin E, which is fat soluble, vitamin C is soluble in water. It occurs in all fruits and vegetables and in animal products that have not been overcooked. All plants make ascorbic acid, as do almost all animals. The very short list of higher organisms that do not produce vitamin C includes the Indian fruit bat, guinea pigs, monkeys, gorillas and human beings. Everything else produces ascorbic acid and lots of it.

Until fairly recently, it was believed that insects do not produce ascorbic acid. I remember being at a meeting of the American Aging Association where a scientist named James Fleming reported that fruit flies make vitamin C. I made the comment that the world's leading authority on the biochemistry of ascorbic acid claimed that insects do not produce vitamin C and that I had found no vitamin C in my fruit flies. I went home thinking that Dr. Fleming was too good a scientist to be wrong.

Insects, unlike animals, have very high levels of the metal ions of copper and iron. Measuring ascorbic acid using the usual techniques without controlling these metal ions, leads to nothing because these metal ions degrade the vitamin C before you can analyze it. Once we figured out how to control this problem, the result was that Drosophila make a large quantity of vitamin C. Fleming was right, and the authoritative expert was wrong.

This result helped to explain why dietary vitamin C has little or no influence on the life span of Drosophila. The amount naturally produced in these insects is so great that dietary supplementation probably does not greatly affect their cellular levels. It is curious that dietary supplementation of ascorbic acid does very little to the life span of fruit flies and even shortens it at high concentrations. Maybe, it is the high metal ion concentrations in these creatures that leads to the destruction of vitamin C. See Chapter 11 for more details.

When the amount of ascorbic acid in the Swedish C and Oregon R strains of Drosophila is compared, there is a big difference. Surprisingly, the amount of vitamin C in the rapidly aging Swedish C flies is about 50% higher than in the Oregon R strain that lives about twice as long. The decline with aging is also greatly different with the Oregon R strain showing a 20% decline with aging, whereas the Swedish C strain experiences a very large 70% drop with increasing age. Perhaps, the reason the Swedish C strain ages more rapidly is due to this large decline in the amount of vitamin C with increasing age. Feeding vitamin C to Swedish C flies for their entire lifetime at a wide range of concentrations, however, failed to produce any significant change in life span. We found the same result for the Oregon R strain. Overall, it seems that very large changes occur in vitamin C concentrations with aging in Drosophila. Dietary supplements are, however, unable to improve life span.

Perhaps we were looking at the wrong organisms. Maybe a better way of looking at the problem would be to examine biological systems that do not produce vitamin C. Very few organisms do not produce vitamin C. One is the Indian fruit bat. They, of course, eat fruit which is a rich source of ascorbic acid. Guinea pigs are strictly vegetarians, and gorillas eat as much as 25 pounds of fruit and vegetables a day. These organisms really do not need to synthesize vitamin C. Human beings also do not make ascorbic acid. Compared to other organisms of a similar size, humans live a long time. Many human diets are very low in vitamin C. This poses a problem. The average gorilla of comparable size to a human consumes 9,000 milligrams of vitamin C per day, while the average human consumes less than a hundredth of that amount. If vitamin C is an important anitoxidant in the aging process, then why do gorillas not live a lot longer than humans? In fact, they age faster than humans.

One of the best biological systems for testing the effect of vitamin C on life span is the guinea pig since it produces no ascorbic acid and is dependent on only dietary sources. The surprising result is that a 1% concentration of vitamin C in the drinking water of guinea pigs produces a reduction in the life span of about 10%. This is in conflict with everything we know about vitamin C. It increases the excretion of cholesterol, reduces the toxicity of metal ions and is absolutely necessary for the synthesis of many important proteins, including the connective tissue protein collagen that makes up much of our tendons and bone. This is a very serious result that must be explained. It clearly suggests that high dietary intakes of vitamin C can be life shortening.

However, this life shortening result was not entirely our experience. When we fed a 1% concentration of ascorbic acid to mice in their drinking water there was a highly significant increase in their median life span of more than 20%. We used distilled water that contains no metal ions. Perhaps this was the reason for our favorable result. We began the feeding at 37 days of age and continued until the mice died of old age. This result was remarkable since mice naturally produce lots of vitamin C. It should be said that the increase in the maximum life span was very little. The maximum survival time for the control group was 965 days versus 993 days for the 1% ascorbic acid group. This experiment suggests that high doses of vitamin C may improve average survival, but they are not the answer to the problem of aging since the maximum life span was essentially unchanged. Whether or not high doses of ascorbic acid are toxic is not known, but this should be considered as a possibility.

There is no doubt, however, that ascorbic acid is of great benefit to humans. Vitamin C supplements do improve the quality of life by reducing certain disease states. Scurvy is a terrible disease that used to kill millions of people worldwide. Today it is seldom seen. This is primarily due to changes in diet. Christopher Columbus, in my opinion, was probably the greatest contributor to medical science of all time when he brought back the potato from the New World to Europe. He, of course, had no idea what he was doing. At that time no one knew about vitamin C, but suddenly the enormous number of people dying during the winter months began to decline thoughout Europe. Millions of people were saved from painful death by the introduction of this plant into the agricultural system of the European continent. More people in Europe died of scurvy than in all their wars combined. Yet, it all ended except on ships at sea, and no one knew why. The British finally figured this out and began to issue lime juice, a good source of ascorbic acid, to their sailors on naval voyages. This is why British sailors are sometimes referred to as Limeys. Again, they really did not know the basis for doing this, except that it worked.

The health benefits of supplemental dietary ascorbic acid have been widely reported. Low dose dietary supplements of vitamin C have been used to moderately lower the mortality rate of humans. It also lowers serum cholesterol and triglycerides in humans. It has been proposed that ascorbic acid can prevent or even reverse atherosclerosis. The mechanism for this appears to be increased conversion of cholesterol to bile acids with subsequent fecal excretion of cholesterol. Ascorbic acid can decease the incidence and delay the onset of skin cancer in mice. It is also toxic to cancerous melanoma cells and can inhibit the mutations induced by known carcinogens. Improvement of cell-mediated immunity (an important part of the immune system which protects us from infections and cancer) in old humans has been reported after injections of vitamin C.

The decline of ascorbic acid concentrations in biological tissues with aging has been reported for organisms as diverse as cattle, rabbits, guinea pigs, rats, mice, fruit flies and man, but its role in the aging process still remains unclear. So far, attempts to improve the life span with ascorbic acid in model experimental systems have been disappointing. Ascorbic acid changes appear to be more of an indicator of the biochemical aging process rather than an effective tool for intervention. These changes tell us that something is wrong, but adding extra dietary vitamin C does not seem to solve the problem of aging. Other more important factors must be involved.

The major question is how much vitamin C is needed? Some very important scientists such as Linus Pauling, who incidently won the Nobel Prize two times (one for chemistry and the other for peace) and was a professor who gave lectures to me in graduate school, have forcefully made the case that we need the same amount of vitamin C as a gorilla. That would be 9,000 milligrams a day. Anyone who has tried to consume this amount of vitamin C per day knows that it produces intestinal distress. Linus Pauling said that was a good thing; it cleaned you out. Perhaps he was right. It is very difficult to argue with a two-time Nobel Prize winner.

One, however, cannot ignore the guinea pig result of a reduced life span of 10% with high dietary intakes of vitamin C. The same is true of fruit flies at high dietary concentrations of ascorbic acid with the reduction being between 8 and 13%. The mouse experiment showing an increase of 20% in median life span with a high dietary intake of vitamin C, however, throws this into doubt. Are vitamin C supplements really necessary, and are high dietary intakes toxic? There are no real answers to these questions. Based on the evidence, every person must make his or her own decision.

Humans do not produce vitamin C, and yet they live a long time. Perhaps there is an unkown reason for this. There is a possibility that excessive intakes of ascorbic acid could be unhealthy. My personal guess (and it is just that) after 30 years of considering this problem is that a moderate supplement of vitamin C would improve health and extend the average, but possibly not the maximum human life span. What is moderate? Perhaps 100 to 500 milligrams per day would be adequate without causing toxicity. I take 500 milligrams per day, but I must say that this may be too much, or maybe too little. One would think that a definite recommendation could be made after all these years and hundreds of research papers. The fact is that the key experiments have not been done.

Vitamin C and vitamin E are the best known of all the natural antioxidants. Neither one of them, however, is the answer to the problem of biological aging. There are many other antioxidants, both synthetic and natural. Some of these have been studied for their effect on the aging process. Many of these are approved food additives.

BHT and BHA

Two commonly used food antioxidants have been examined for their effect on the life span of rodents. Butylated hydroxy toluene (BHT) and butylated hydroxy anusol (BHA) have long been used as food additives in order to prevent the peroxidation of fats. They used to be added to almost all potato chips and many baked products and frozen foods. They are still used today to a lesser extent.

Some manufacturers add them, not to the food itself, but to the packaging material. This prevents oxygen from getting to the product. When fed to mice, BHT did improve the average life span to a limited extent but did not increase the maximum life span. BHT and BHA greatly improve the life span of a potato chip, but they seem to do very little for prolonging the maximum life span of living biological organisms.

NDGA

There are numerous other antioxidant food additives, and almost all those tested for life span improvement in Drosophila had no effect. Just to mention one, we looked at propyll gallate, and it had no effect on life span. However, one food additive, which did give a considerable extension of life span, was nordihydroguaiaretic acid or NDGA. According to the Merk Index, which is a marvelous collection of information on all kinds of chemical compounds, NDGA is used as an antioxidant for fats and oils. "Lard containing 0.01% NDGA stored at room temperature for 19 months in diffused daylight showed no appreciable rancidity or color change. NDGA is a natural product occurring in the resinous exudates of many plants." One of these plants is the creosote bush which grows in the desert of Arizona, and is one of the longest living organisms on earth.

We examined NDGA in adult fruit flies and found no improvement in life span. However, when fly eggs were placed on NDGA and allowed to develop, there was a 20% increase in life span for the emerging adults in both the average and maximum life span. At the time, I thought that this was an interesting, but not very useful, result. If the result were extended to humans, one would have to recommend that pregnant women take NDGA in order for their children to enjoy a longer life span. There is also the possibility of fetal toxicity. In addition, NDGA is no longer an approved food additive. We dismissed the result and never published it. This was, of course, a mistake. Twelve years later a paper appeared (Richie, Mills and Lang, Proceedings of the Society for Experimental Biology and Medicine, vol. 183, pp. 81-85, 1986).

Using the mosquito as a model aging system, the authors reported greatly increased survival for both male and female mosquitoes when they were given very low concentrations of NDGA as young adults. The average adult life span was extended by 42 to 64%. The maximum life span was also increased but to a lesser extent. Mosquitoes, like fruit flies, have a very short life span of a month or two. When NDGA was given to older adult mosquitoes, there was no effect. The interesting thing is that it worked on young adult mosquitoes, whereas we found no change at all when young or old adult fruit flies were given NDGA.

There is also limited support for this result in other systems, including a fungus, which has been used for genetic research, and the laboratory rat. This certainly seems like an important topic for further research. To my knowledge, these interesting findings have, unfortunately, not been followed up. Perhaps the reason is that NDGA has no effect on older organisms. We can reasonably conclude that NDGA seems to prevent the initiation of lipid peroxidation. It is also known that NDGA inhibits the formation of the aging pigment, lipofuscin, but apparently, after this process has begun, NDGA is unable to influence it.

Methionine
Another interesting antioxidant is methionine. Methionine is a water-soluble essential amino acid needed for nor-

mal metabolic function. It is one of the few amino acids to contain sulfur. It has been suggested that methionine may also function as an inhibitor of the aging process based upon its ability to inhibit superoxide radical production from metabolically active human mitochondria. This, of course, assumes that the free radical theory of aging is correct.

Other observations suggest a protective role for methionine in the aging process. Formation of the age pigments (Chapter 7) is inhibited by methionine. A combination of vitamin E and methionine induces higher levels of the enzyme glutathione peroxidase which is involved in the removal of lipid peroxides from tissues. Methionine has been reported to decrease the lipid peroxidation induced by the chemical, carbon tetrachloride. The eye lenses of older humans show oxidation of methionine in membrane fractions with extensive oxidation of methionine occurring in the regions of the lenses containing cataracts. This indicates that methionine may protect the eyes from the damaging effects of atmospheric oxygen. Methionine also protects rats from carcinogenisis induced by both natural and synthetic cancer causing compounds.

A diet low in methionine leads to elevated cholesterol levels and atherosclerosis in rats, mice and monkeys, but, for some mysterious reason, it has no effect on pigs. Older rats absorb methionine at a slower rate, and it disappears more rapidly from the tissues of old rats than from young rats. All of this evidence suggests that dietary methionine might influence the aging process.

In our first experiment with methionine, we did find a large increase in the life span of middle aged Drosophila when they were fed extra methionine. We were never able to repeat this result. This illustrates the importance of repeating experiments. It is essential to repeat experiments to insure their validity. We even tried to see if the oxidation products of methionine were responsible for this result since the methionine we used was from an old bottle. Methionine sulfone and methionine sulfoxide, oxidation products of methionine, did not change the survival times. We tried rearing the flies on methionine and feeding it to flies of young, middle and old age. None of these attempts produced an improvement in life span, and the higher doses actually significantly reduced survival. I concluded that dietary methionine was unable to improve the life span of Drosophila of any age and could even considerably reduce it at higher doses.

Feeding combinations of methionine and vitamin E, which as noted above is supposed to increase the activity levels of the antioxidant enzyme glutathione peroxidase, also failed to improve life span of Drosophila.

The methionine story was so compelling, however, that I went ahead with studies in mice. The result was overwhelmingly disappointing. When young mice, beginning at 42 days of age, were given extra methionine in their diet, there was a considerable reduction of 17% in their life span. Old mice given extra methionine, beginning at 581 days of age, showed no change in life span. As one can see, the whole methionine theory of aging fell apart.

What was the reason for this result? I failed to consider all the negative research concerning methionine. Humans cannot produce methionine just as they cannot produce vitamin C. It is an absolutely essential nutrient in the diet. One would, therefore, tend to consider it to be of benefit in general. There is some question about whether or not high concentrations of vitamin C can be toxic, but there is little doubt that high doses of methionine can be a problem. It has been observed that chickens fed 1% or more of methionine have a decreased rate of weight gain, lower blood hemoglobin and increased spleen iron levels. Methionine also produces nuclear and nucleolar lesions at the center of the cell.

Methionine has been described by one researcher as being the most toxic of all the nutritionally important amino acids. There is a great species difference in the sensitivity to methionine. Rats can consume 1.5% methionine in their diets with no apparent decrease in the rate of growth, whereas guinea pigs, rabbits and man are less tolerant of methionine. A dose of 10millimoles/kg/day results in inanition (lack of movement) and death within 65 hours for guinea pigs. A single small intravenous dose of 2.8 millimoles/kg of methionine to humans produces nausea, vomiting, low blood pressure, abnormal heartbeats, fever, disorientation and liver dysfunction. Dietary sources of methionine are large for most people. Milk protein from cows contains 2.6% methionine, and chicken egg albumen protein contains 4.6% methionine. That is a lot of methionine. Dietary deficiency, therefore, seems unlikely. The question of how much is toxic to humans is another matter that is open to consideration.

All of these antioxidants, including methionine, have been mixed together and fed to mice for most of their life span. The result was that the age-related accumulation of fluorescent age pigments was reduced, but there was no improvement in life span. The investigators tried different concentrations and combinations of vitamin E, BHT, vitamin C, methionine and selenium (which is at the active site of the enzyme glutathione peroxidase). It seems that even combinations of both the fat and water-soluble antioxidants do essentially nothing for overall life span. This was, of course, disappointing to those involved.

Uric Acid

Uric acid has been shown to be a powerful antioxidant. It also is a scavenger of the atmospheric pollutant ozone, singlet oxygen (discussed in Chapter 9) and free radicals. Uric acid is formed by the action of the enzyme xanthine oxidase on the purine molecule xanthine (a molecule generally found in certain foods such as liver and other organ meats). Uric acid is, therefore, referred to as the end product of purine metabolism in man and the other primates. Proctor, in 1970, (Nature, vol. 228, p. 868) made the very interesting proposal that uric acid serves to replace vitamin C as an antioxidant in organisms unable to synthesize vitamin C. This idea is supported by the fact that the amount of uric acid in human blood very greatly exceeds that of vitamin C.

I looked at how uric acid levels change with age in fruit flies and found a 59% decline with age. The molecule from which uric acid is formed, xanthine, also showed a considerable decline of 75% as the flies aged. Feeding xanthine or uric acid to fruit flies, however, gave no improvement in life span. It should be pointed out that the solubility in water of these two purines is very low so we were unable to add high dietary levels.

Humans normally have very high concentrations of uric acid in the blood. People with higher than normal uric acid levels sometimes develop the painful disease of gout due to crystals of uric acid forming in the joints. Any attempt to increase uric acid in humans could, therefore, be risky. Uric acid levels are elevated in people who drink alcohol and in those who take diuretics for the control of high blood pressure. It can be concluded that the large amount of uric acid in human blood is probably there for a good reason in terms of antioxidant defense, but to increase it even more would probably be undesirable.

Chapter 9

SINGLET OXYGEN AND BETA-CAROTENE

Beta-carotene is another antioxidant that has received a great deal of attention. It is a natural product produced by all plants. Almost all animals and humans can convert it to vitamin A. Unlike vitamin A, it is not toxic when consumed at moderately elevated concentrations. It is fat-soluble, and in order for absorption to occur, there must be a certain amount of fat in the diet. This accounts for the fact that many thousands of children used to go blind for lack of vitamin A in their diets in spite of the fact that they ate a high beta-carotene diet consisting primarily of vegetables and grains. They had plenty of beta-carotene in their diets but could not absorb it because of a lack of fat in their foods. The result was vitamin A deficiency and resultant blindness.

The World Health Organization and other agencies have distributed very high dose vitamin A pills to some children in an effort to correct this problem. A small amount of cooking oil would accomplish the same purpose without the possibility of vitamin A toxicity.

Carotenoids

Carrots and sweet potatoes have a high content of beta-carotene, and this is why they are orange in color. The greatest dietary source of beta-carotene is the red palm oil that is consumed in western Nigeria. It is of interest that claims of enormous life spans of 150 years and more have been reported in this area of Nigeria. I have attempted to verify these claims without success, but it is something that should be investigated. The butter from range fed cows is yellow. The butter from their milk is yellow in color due to the beta-carotene in the grasses and other plant foods they eat. Grain fed cows have white colored butter.

Beta-carotene is just one of the many plant carotenoids which lend color to various plants. The real function of these carotenoids, however, is to protect plants from light and the highly oxidizing environment in which they live. A great deal of research has been done on the carotenoids. The chemistry on this subject is very impressive. Not only have the structures of these very large and complicated molecules been established, but their mode of chemical action has also been worked out. One of the results of this brilliant research has been the discovery that beta-carotene reacts with singlet oxygen.

Singlet oxygen

Instead of free radicals with unpaired electrons, suppose that highly energetic molecules, such as singlet oxygen, are the real problem in aging. Singlet oxygen is chemically the same as ordinary oxygen which is triplet oxygen. Both are composed of two atoms of oxygen. The difference is that the electrons in singlet oxygen are more energetic than those of triplet oxygen. The chemists say that they are at a higher electronic energy level. One could say that singlet oxygen is similar to a person who has consumed too much caffeine and sugar. It is more energetic and reactive. Singlet oxygen will do all the things triplet oxygen will do, but more readily and at a faster rate. From the point of view of aging research, this seems to be of obvious interest.

Almost all molecules can be induced to a higher energy state. This is usually done with exposure to visible light. If that does not work, then they can be exposed to more powerful ultraviolet light. Higher energy states of the electrons in molecules can also be induced chemically. Mixing ordinary household bleach with readily available drug store hydrogen peroxide produces singlet oxygen. The bubbles coming off are singlet oxygen. There is nothing really unique or complicated about producing it. It is, however, not a good idea to do this and yet our bodies do it every day.

The primary killing agent used by our immune system is believed to be sodium hypochlorite (bleach). The cells of the immune system secrete it whenever we are infected with viruses, bacteria, fungus or even cancer cells. Long before man invented bleach, cells of the immume system have produced this substance in abundance. The body also produces large quantities of hydrogen peroxide every day as a result of normal metabolism. Therefore, it seems obvious that at least some singlet oxygen must be generated in humans and most other organisms as a result of the reaction of these two molecules. It is even possible that singlet oxygen itself is the primary killing agent of the immune system and not sodium hypochlorite (bleach). There is, however, no scientific evidence for this proposal.

There are many other ways of producing singlet oxygen. It is believed that the major pathway for singlet oxygen production in biological systems is the reaction between two superoxide radicals. These are "free" free radicals and most, but not all, are normally removed by the enzyme superoxide dismutase (SOD). They are involved in the process of lipid peroxidation, and possibly the reason for this is that they combine to produce singlet oxygen. It is known that singlet oxygen reacts at a rate 1,500 times faster than the ordinary triplet oxygen which we breathe. When exposed to the fat, methyl linoleate, several laboratories have shown that singlet oxygen can cause oxidation of lipids.

Cancer and diet
It is possible that singlet oxygen is the reason for aging or at least part of the reason. It can react with almost every molecule in the body. Since beta-carotene eliminates singlet oxygen, one would expect that beta-carotene would prolong life span or reduce the incidence of important disease states. The great interest in beta-carotene began with an epidemiological study which lasted for 19 years and showed that a large group of 1,954 men, 40-55 years of age at the beginning of the study, working for the Western Electric Company in Chicago, had an enormous decrease in the rate of lung cancer for smokers who consumed a diet high in vegetables, soup and fruit.

What was especially interesting was that even long time cigarette smokers of 30 years or more had almost no increase in the rate of lung cancer when they consumed these dietary factors. The guess was that this was due to the beta-carotene content of the vegetables, fruits and soups. An additional 8 studies confirmed these results and extended them to other forms of cancer for both smokers and non-smokers. Many nutrition experts, at this time, said that smokers should take beta-carotene pills, or at least eat lots of carrots since they contain large amounts of beta-carotene. At this point no one knew what effect beta-carotene supplementation had on cancer or any other disease state, including cardiovascular disease. There were many predictions but no real evidence.

Beta-carotene and aging

Before this result was reported, I had examined the effect of dietary beta-carotene on the life span of Drosophila. Since beta-carotene is not water-soluble, we had to mix it with either alcohol or corn oil. We tried every possible age group of flies and concentration of beta-carotene. The flies turned orange, so I knew that it was being absorbed. We found a prolongation of life for beta-carotene dissolved in corn oil, but corn oil itself gave almost the same increase in life span. Some concentrations of beta-carotene actually reduced life span. We did, however, find a 9% increase in life span when flies were reared and maintained as adults on medium containing beta-carotene dissolved in 5% alcohol. Otherwise, there was no improvement in life span.

Since the flies turned orange in color, I knew that the beta-carotene was being incorporated into the fats of the fly cells. It was possible, however, that the damage being done by singlet oxygen was not to fats but to the water-soluble areas of the cell. We, therefore, tried a water-soluble compound known to react with singlet oxygen, 1,4-diazabicyclo (2.2.2) octane or DABCO for short. This compound did improve life span by a mere 7% at low concentrations but reduced life span at a higher concentration by 12%.

Sodium hypochlorite, or common bleach, is supposed to react with the hydrogen peroxide in the cell to produce singlet oxygen and should, therefore, greatly reduce life span. The surprising result was that feeding adult flies bleach at a concentration of 0.001% in their food actually improved their life spans by 11%. Only when we went to a concentration of 0.01% was there a reduction of 16% in the life span. The disturbing fact was that sodium hypochlorite prolonged life span more than either beta-carotene or DABCO (the water soluble molecule which reacts with singlet oxygen). My guess is that bleach improves immune function and that this is why flies live longer on diets containing low concentrations of bleach.

This result threw the whole idea of the singlet oxygen theory into serious question. I had only one idea left: bleach needs hydrogen peroxide to produce singlet oxygen. It turns out that catalase, the enzyme that normally removes hydrogen peroxide, can be made inactive with 2-amino-1,2,4-triazole, or 3AT for short. When we fed 0.05% 3AT to flies, there was a 11% reduction in life span. This indicated that excessive hydrogen peroxide by itself could reduce life span. When we added sodium hypochlorite to the diet containing 3AT, there was an even greater reduction in life span of 25%. This seemed to indicate that singlet oxygen could reduce life span. Just to be sure, we added beta-carotene and DABCO to the mix to see if we could reverse the effect. These two compounds did partially, but not completely reverse the life shortening effect. Overall, the results seemed to say that singlet oxygen could reduce life span if produced artificially, but that the low level of naturally generated singlet oxygen has little or no effect on life span.

The increase in life span with bleach was shocking. Bleach is a powerful oxidizing agent and yet it improves life span. This suggests that singlet oxygen and bleach are very active and necessary components of immune function that override any mechanism of the basic aging process. This is another theory of aging. The loss of immune function with aging is considerable and is of great importance. The reason for this remains unknown, but it is an important aspect of the aging process.

While the results with Drosophila and beta-carotene did not turn out to be very promising, I felt that the evidence from other sources was so convincing for a role of singlet oxygen and beta-carotene in the aging process that an experiment with mice was justified. The abstract of our published paper summarizes our findings:

"Feeding 0.5% beta-carotene in the diet for life beginning at 29 days of age improved the average life span of C57BL/6J male mice by 5.0% but decreased the life span of mice started at 608 days of age by 11.5%. Neither difference, however, proved to be statistically significant. Feeding beta-carotene increased the concentration of beta-carotene in the serum by 60% but did not change the beta-carotene content of heart, liver or kidney. We can conclude that singlet oxygen, which is very efficiently quenched by beta-carotene, is an important factor in senescence only if it is produced at organ sites not accessible to serum beta-carotene. Since beta-carotene feeding is not a useful means for increasing tissue concentrations of beta-carotene, other more sophisticated means must be developed for accomplishing this purpose. It is also clear that, while dietary beta-carotene is not an effective means for prolonging life span, it is nontoxic when fed continuously at high concentrations."

Beta-carotene and cancer
It turns out that this last observation was tragically not true for humans. As we will see in the next few paragraphs, excess beta-carotene can be toxic to humans.

Once again, the fruit fly results were similar to those

for mice. We had to conclude that beta-carotene had essentially no influence on life span in two totally different biological systems and that the singlet oxygen theory of aging was probably wrong. The interest in beta-carotene, however, continued based on several reports of its apparent anticancer properties, but not all research supported this idea. One laboratory reported no effect while another found that beta-carotene greatly increased the number of tumors produced when applied with the tumor promoter, croton resin. You might, therefore, predict that high dietary intakes of beta-carotene, which remove singlet oxygen, would actually increase the rate of cancer and perhaps other diseases.

This is exactly what was found when the National Institutes of Health funded a study where thousands of physicians agreed to take low dose beta-carotene pills. Half of them consumed extra beta-carotene, and the other half did not. The result was that the incidence of cancer actually increased in the group taking the beta-carotene. What was especially disturbing was that those who smoked had a much higher incidence of lung cancer. This pretty much killed the beta-carotene story, but one has to wonder what was in the soup, fruit and vegetables, in the epidemiological study discussed above, that the Western Electric Company employees ate which reduced their incidence of lung cancer by ten fold.

It certainly was not beta-carotene, much to the embarrassment of the health food experts, the scientists and the physicians (especially those who took the beta-carotene supplements and developed lung cancer). Some scientists had cautioned that another factor other than beta-carotene might be involved. Perhaps it was lycopene, which is another plant-derived antioxidant, and is found in high concentrations in tomatoes. Much to my regret, I did not do any aging research on this very important molecule.

It is of interest that dietary supplements of beta-carotene reduce the blood levels of vitamin E in rats by 50%. In chickens, high dietary beta-carotene reduced liver vitamin E by threefold. In humans, feeding beta-carotene for six months lowered plasma vitamin E levels by 40%. This may have been the reason that beta-carotene pills did nothing for the thousands of physicians taking it and increased the incidence of lung cancer. Recently, it has been shown that smokers taking beta-carotene pills are also at greater risk for colorectal, thyroid, ovarian and cervical cancers, as well.

Errors of judgment
Several conclusions can be drawn from all of these studies. One is that the singlet oxygen theory of aging is probably wrong. Another is that extra dietary beta-carotene is of no use for increasing life span and actually increases the risk of cancer, especially lung cancer. Dietary vegetables, soup and fruit contain something that greatly reduces the incidence of lung and other cancers, especially if you are a smoker, but it is not beta-carotene.

The mistake in all of this research was to assume that a single factor, beta-carotene, was involved. It certainly made a lot of sense, but it was wrong.

Chapter 10

WHOLE FOOD ANTIOXIDANTS

The lesson to be learned from the beta-carotene experience, to me, seems to be that we should take positive results, no matter how ill defined they are, and work backwards. In the Western Electric Co. study, it was either the vegetables, the soup or the fruit. It could have been plant or animal products depending on what was in the soups. Instead of beta-carotene, we should have been looking at extracts of these food sources. After finding favorable results we then could have worked backwards to find the source. As it was, we guessed the source and ended up with little or nothing to show for our efforts.

Whole foods and life span
One of my favorite research papers is: The Antioxidant Properties of Spices in Foods, by Chipault, Mizuno and Lundberg (Food Technology, vol. 10, 1956). The authors looked at the antioxidant properties of allspice, cardamom, cassia, cinnamon, cloves, ginger, mace, mustard, nutmeg, oregano, black pepper, white pepper, rosemary, sage, savory, thyme, turmeric, and spice mixture (sage, black pepper, red pepper and salt). This paper convinced

me that whole food antioxidants are more powerful than single chemical antioxidants.

Among all the spices they tested for anitoxidant activity, cloves and sage were high, but highest of all was oregano. We tried oregano on our fruit flies. We prepared an alcohol extract of oregano leaves. A one to twenty dilution of this extract reduced life span by 15%. This was a shock. It was toxic. We tried again with greater dilutions of the alcohol extract. A one to 100 dilution produced no change in life span. A one to 200 dilution increased life span by 7%, and a dilution of one to 300 increased life span by 12%. Apparently, there is a reason why insects do not attack most spices. At full concentration they seem to be toxic. In the case of oregano, low concentrations seem to be of benefit to life span, but high concentrations have a negative effect.

Strawberries have been reported to have antioxidant activity. We tried a homogenate (ground up solution) of frozen strawberries. At a full concentration there was no change in life span, but when the homogenate was diluted by one to ten there was a 16.4% increase in the average life span. It is peculiar that, for both strawberries and oregano that high concentrations have no effect, or a negative effect on life span.

Spinach and blueberries have recently been reported to have high antioxidant activity, but we did not try them. As far as I know, no one has looked at their effect on life span.

Redwood trees live for thousands of years. It makes sense that they must have something unique. A small amount of interesting chemistry has been done on this subject. We bought a piece of redwood at a lumberyard and ground it and made an extract of the sawdust. At full concentration it extended the life span of Drosophila by an impressive 23.5%. Both the average and maximum life span were extended. Lower concentrations had no influence on life span. A few people thought that I had lost my mind at this point. The point is that it worked. I never published this result or extended the research. Very high concentrations of the extract did not prolong life span, as was the case for oregano and strawberries. It should be emphasized that we did this experiment only once and that it is why it was not published. The potential for human health and life span using redwood extracts could be enormous.

The root of the ginseng plant is supposed to do all kinds of favorable things, including increasing energy and prolonging life span. There was no improvement in life span and reduced life span at higher concentrations. I published this result. A gentleman from Wisconsin wrote to me saying that I had used the wrong kind of ginseng and sent a sample of his own ginseng grown in Wisconsin. His sample also failed to improve the survival of fruit flies. It also reduced the life span by essentially the same amount, 15%, at the higher concentrations. I also published this result since there was such a great interest in this subject.

English evening primrose oil is also a product that has a considerable reputation for antiaging properties. It had no effect on life span and greatly reduced life span of Drosophila at high concentrations.

Tea
Tea (Camellia sinensis) is a beverage consumed in most parts of the world. Originally it was used as a medicine. Because of its great economic importance, there is a considerable literature on its history, agriculture and chemistry. We found that water extracts of it, prepared just as you would normally brew tea, prolonged both the average

and maximum life span of fruit flies. Different brands of commercially available tea gave various increases in life span. Black tea gave the greatest improvement ranging from 16.7% to 21.4%. Even greatly diluted black tea improved life span. High concentrations of tea extract were not as beneficial as the more dilute concentrations. Green tea gave an improvement of 8.3%. Decaffeinated tea had no influence on life span. Apparently the decaffeinating process removes the life-prolonging factor or factors.

If you are interested in tea, a very informative article to read is, "Tea: The Plant and Its Manufacture; Chemistry and Consumption of the Beverage," by Harold Graham, Progress in Clinical and Biological Research, vol. 158, pp. 29-73, 1984.

Since there is so much known about the chemical composition of tea, we tried to narrow down the causative factor for survival improvement in tea. Caffeine is the most obvious factor. A wide range of concentrations of caffeine had no influence, either negatively or positively, on life span. Another component of tea is catechin. It also had no influence on life span. Epicatechin also failed to prolong the survival of Drosophila. Catechin and epicatechin are the major polyphenol (long molecules that have phenol at the basis of their structure) components of tea. These molecules are believed to be at the basis of the favorable effects of tea on human health. We found that they had no effect at all on the life span of Drosophila. Tannic acid, which we were sure would make a difference in survival, also failed to change life span. Theobromine and theophyline, other important components of tea, did nothing. Gallic acid, which is yet another major component of tea, also gave no improvement in life span.

Thus, our efforts to come up with a unique chemical component of tea, responsible for its life prolonging property, failed. Tea contains at least a hundred different known molecules. As a logical conclusion, perhaps, it is a combination of these molecules, rather than a particular one, that is responsible for the beneficial effect of tea. We published our results on tea in 1993, and they were more or less greeted with a yawn. Perhaps this was because the research was done with fruit flies.

Since then, an amazing amount of evidence indicates that drinking tea is of great benefit to humans. Most of this has been reported in the popular press. Extracts of tea, especially green tea, have been shown to inhibit cancers. Some of the components of green tea have been isolated and used to kill cancer cells. A Dutch study showed that drinking one to three cups of regular black tea daily cut the risk of severe arteriosclerosis (artery clogging) by 46% in older men and women. Four cups of tea a day lowered the risk by 69%. Another recent study showed that humans who drink tea have immune cells that responded five times faster to germs than do the blood cells of coffee drinkers. This is believed to be due to the L-theanine in tea according to the researchers. This may account for the numerous biochemical studies showing anticancer properties of molecules extracted from tea. There is even a report that tea prevents bad breath.

These are not trivial results. At least some of the many benefits resulting from drinking tea may be due to its effect on the aging process. Reducing the rate of aging will most certainly reduce the incidence of both cancer and cardiovascular disease, since they are so clearly related to the aging process in humans. One could take the opposite approach by trying to reduce cancer and cardiovascular disease in order to reduce the rate of aging. So far this approach, which has received considerable attention, has produced some success. Tea offers an amazing and economical approach to aging and the diseases associated with it. To understand its mechanism of action would seem to be of utmost importance.

There is one negative side effect to excess tea consumption that must be kept in mind. It destroys the B1 vitamin, thiamine. This can lead to the disease beriberi. In northern Thailand, for example, fermented tea leaves are chewed continuously as a stimulant. As many as 21.6% of these people have thiamine deficiency which can result in beriberi. Excessive tea consumption can also lead to anemia in children and young women. I give a possible reason for this in a following chapter. Just like everything else, excess tea consumption can lead to problems.

Chapter 11

IRON

Why does tea prolong the life span of fruit flies? We found that it eliminates the accumulation of iron that normally occurs with aging in Drosophila.

Dietary iron is probably one of the most contentious subjects you can write about. There are those who believe that it is of great benefit and those who think that it can be a poison. There is absolutely no doubt that iron is an essential nutrient. It is needed to make the blood protein hemoglobin and many enzymes in the body including catalase which removes hydrogen peroxide. To make the suggestion that iron overload is a health problem can be difficult. For example, I wrote a grant proposal to the National Institutes of Health where I proposed that excess dietary iron was a major cause of aging. It was rejected.

Dietary iron
The fact is that dietary iron, like most nutrients, is important, but it can also be a source of serious problems. Since there is no known mechanism for the excretion of excess iron, it is possible that iron overload is a risk for adult humans, especially men and postmenopausal women.

Growing children and premenopausal women, in contrast, rarely experience iron overload because of growth
and blood loss. This is indicated by the fact that serum
ferritin, which is the best indicator of total body iron content, increases from 30 micrograms per liter at age 15 to
more than 90 micrograms/liter at age 30 in males. After
age 30 serum ferritin continues to increase gradually in
most males. My own serum ferritin reached a value of
560. I thought that this was too high. I donated blood 3
times, and it is now 80. In females, however, the increase
in serum ferritin is delayed until after age 40. This observation has prompted a physician and scientist named Je-
rome Sullivan to hypothesize that the greater incidence of
heart disease in older men and postmenopausal women
compared with the incidence in premenopausal women
and young men results from the higher levels of stored
iron in these two age groups.

Heart disease, cancer and osteoporosis
In support of Dr. Sullivan's idea is the observation that
myocardial failure is a prominent feature in iron overload
diseases such as hemochromatosis and thalassemia major.
Other disease states are also associated with excess iron.
In patients with iron overload, functional impairment of
the immume system cells, monocytes and granulocytes,
has been reported. Among iron miners in France, the incidence of lung cancer is more than 3 times the number
in the French male population in the same age group.
Elevated iron may be involved in rheumatoid arthritis
disease. In the brain disease, neuronal ceroid lipofuscino-
ses, higher concentrations of iron have been found in the
cerebrospinal fluid of patients. Many cases of severe spinal osteoporosis (bone loss disease) have been found in
iron overloaded young South African males (yes, males,

not females). These young men drink a lot of beer fermented in iron pots, and this high-iron beer is believed to be the reason for their osteoporosis. The very high incidence of osteoporosis in elderly women could also be due to the age-related iron accumulation that occurs after menopause.

It is also known that iron carbohydrate complexes (iron combined mostly with sugars) are carcinogenic, and that iron sulfate is mutagenic (causes DNA changes) in bacteria. Dr. Weinberg (Physiol. Rev., vol. 64, pp. 65-102, 1984) concludes in his review of the literature on iron that "procedures for preventing the accumulation of excess iron in human or animal populations would be expected to lower the incidence not only of infection and neoplasia (cancer) but also of chronic cardiac failure."

The mechanism for tissue damage in iron overload conditions is not firmly established, but it may involve free radical mechanisms where the two forms of ionic iron, ferrous and ferric ions, act as catalysts. Since catalysts accelerate the rate of any chemical reaction, this would explain why relatively small changes in iron concentrations might be responsible for extensive tissue damage.

The Fenton reaction, where iron ions react with hydrogen peroxide to produce the dangerous hydoxyl free radical, is widely believed to be a major source of damaging hydroxyl radicals. Other reactions, including iron-catalyzed lipid peroxidation, have also been proposed. Tissue damage or cell death might occur if cellular membranes are made more permeable by iron catalyzed oxidation reactions. This could lead to an influx of excess calcium into the cell, a process that is known to occur with aging. The result would be hardening of the arteries and calcium accumulation in other soft tissues, including the brain.

Increased iron concentrations have been found to occur

with aging in human tissues with the greatest increase occurring in the main blood vessel coming from the heart, the aorta. It has also been established in other studies that large increases in calcium occur in aortic and femoral arteries during human aging. Clogged femoral arteries can lead to considerable pain in the legs. In diseased aortas, calcium and iron concentrations are considerably elevated not only in the areas containing plaque but also in the areas free of plaque. Other metal ions including manganese, zinc and copper were found to be unchanged.

In other disease states, a close relationship between calcification and iron deposition has been repeatedly observed in various forms of calcium accumulation, referred to as clinical calcinosis. Heavy mineralization with calcium and iron deposits has been reported in tracheal cartilages, kidneys, prostate gland and some arteries in aging rats maintained on a high fat diet. Therefore, it appears that the normal calcification and hardening of tissues including the arteries, which accompanies the aging process, could have its origin in iron-induced damage to the cellular membranes.

The age pigments all contain iron. These particles may represent the site at which metal ions catalyze the production of free radicals and peroxidation reactions. It has also been reported that high accumulations of iron, calcium and aluminum occur in patients with Lou Gehrig's disease and Parkinson's disease.

Added dietary iron
In spite of the overwhelming evidence for the involvement of excess dietary iron in numerous disease states, the use of iron salts in the supplementation of baking flour is mandatory national policy in the United States and several other countries. In addition, the use of di-

etary supplements containing iron is widespread among adults and the aged who have no need for these iron supplements. Although it is known that some children and about 5% of young women require such extra dietary iron supplements due to poor diets, the rest of the population may be exposed to a possible hazard because of the requirement for adding iron to major food sources such as flour and cereals.

Iron is not a harmless addition even to vitamin pills. A chief pharmacist for a major drug store chain has told me that about 5 children a year in the United States die from iron poisoning after eating a whole bottle of those sweet tasting vitamin pills containing added iron made for children. An expert scientist on the subject of iron has said that, of all the metal ions that undergo oxidation reactions "iron has been found to be the most active promoter of lipid peroxidation" both in the test tube and in living organisms. Since the role of iron ions and their complexes is widely acknowledged as a key factor in the promotion of oxidation of a wide range of biological materials, it is not unreasonable to propose that iron excess is involved in the basic aging process.

Mice and fruit flies
We measured the iron concentrations in mice on a normal diet versus age and found that iron increased with age in liver, heart, kidney, brain and bone with bone showing the greatest increase of 207%. Based on this very large increase, I proposed that osteoporosis might be an iron overload disease since all of these male mice developed osteoporosis as they aged. The fact that these were male mice eliminated the possibility that female hormonal changes were involved. This proposal is also consistent with the fact that postmenopausal women, who have

most of the cases of human osteoporosis, experience a very large and sudden increase in total body iron after menopause. Iron also increases with age in both fruit flies and houseflies.

The advantage of studying iron and aging in fruit flies is that we can vary the rate of aging by changing the environmental temperature. If iron accumulation is a real factor in the basic aging process, then it should show different rates of change at different environmental temperatures. This is exactly what we found.

The median life span, which is the age at which half the flies are still alive and half are dead, varies with the environmental temperature. Median life span is almost the same as average life span. At 11 degrees Celsius the median life span is 152 days; at 20 degrees it is 81 days; at 25 degrees it is 62 days, and at 30 degrees it is 25 days. The rate of iron accumulation at these different temperatures varied greatly. On a relative daily basis it ranged from 1.5, 4.2, 7.3 and 15.0 for the four different temperatures with 15.0 being the highest rate at 30 degrees Celsius and 1.5 the lowest rate at 11 degrees Celsius.

Another interesting result is that the rates of iron accumulation with aging in fruit flies at various temperatures, compared to the rates for mice and humans, are directly proportional to average life span in these three different species. Mice accumulate iron more slowly than fruit flies, and humans accumulate iron in the body at a much slower rate than mice. Thus, the faster the rate of iron accumulation, the shorter the life span.

We also found that the large increase in tissue calcium that occurs with aging in fruit flies, and in other aging organisms, including humans, is directly related to the amount of whole body iron. Old flies have very high calcium levels and high iron levels. This supports the idea that iron damage to membranes could be the reason that all of the organs, except bone, accumulate large amounts of calcium during the aging process. This also occurs in certain disease states such as cardiovascular disease.

Tea and vitamin C

Finally, we were able to show that dietary tea completely eliminates the accumulation of iron that normally occurs with aging in fruit flies. The result was that there was usually a 20% increase in the average, median and maximum life span of Drosophila.

As a test of the idea that iron accumulation was the problem which tea corrected, we fed iron to fruit flies in the form of ferrous gluconate (this is what is added to some vitamin pills). This reduced their life span by 26.8%. However, when tea extract was added to the same amount of ferrous gluconate in the diet, the life span was increased by 16.1%. Thus, tea was able to overcome the toxic effect of high iron in the diet to a large extent. This seemed convincing to me and supported the idea that inhibition of iron absorption was probably the principal mechanism for tea prolonging life span. This, of course, does not mean that there were not other reasons for the effect of tea on life span.

Iron accumulation, however, cannot be the whole reason for aging since we found substantial, but not enormous, increases in life span when the normal accumulation of iron was blocked. Whether or not these results extend to humans remains to be seen.

When we added vitamin C to the tea extract, the median life span of Drosophila was increased by 30.9%. There was a similar extension of the maximum life span. We repeated this experiment, and the median life span was extended by 30.6%. Unfortunately, we did not get a maximum life span in this experiment because of yeast growth in the food at 66 days of age, which is quite old for a fruit fly. Flies get stuck in the yeast growth and die, so it ruins the experiment. Even so, at this age 42% of the tea plus ascorbic acid flies were still alive, while only 13% of the control group were still alive. Since the second experiment was not completed due to the yeast infection, we could not consider this for scientific publication. I regret that we did not try again, but there were other pressures at the time. I think that we can say with confidence, however, that tea and vitamin C work together to greatly reduce the rate of aging, at least in fruit flies. We, of course, do not know if this extends to humans, but if it does, it would mean that an increase in life span of at least 20 years could be expected.

Chapter 12

CADMIUM

The biological mechanism for iron absorption is the same as it is for several very toxic metal ions. One of these is cadmium. Next to radioactive metals such as plutonium, cadmium is probably the most toxic of all the metal ions.

The Japanese Ministry of Health and Welfare established in 1968 that Itai-Itai disease, characterized by skeletal deformities and multiple bone fractures, is caused by chronic cadmium poisoning from contaminated shellfish. Two hundred people died of this disease before it was found that cadmium poisoning was the problem. I have a photograph of a young woman who survived the poisoning. It is not something one really wants to look at. Thus, there is little doubt that high levels of cadmium represent a definite environmental and health hazard.

Cadmium and disease

The possible involvement of low levels of cadmium in the development of the major age-dependent degenerative diseases is suggested by the observations that elevated levels of atmospheric cadmium are associated with increased frequency of cardiovascular disease death rates and certain kinds of cancer. At the experimental level small amounts of cadmium have been used successfully to induce high blood pressure in rats and tumors in rat testes.

At the molecular level, cadmium suppresses immune response in rats, decreases the fidelity of DNA synthesis, elevates blood glucose and urea, causes mitochondrial swelling, and respiratory deficiency, uncouples the oxidative phosphorylation process which is necessary for the production of energy molecules, depresses rod receptor potential in the retina, and produces single-strand breaks in DNA. Normal cross-linking of the protein collagen is also completely inhibited by cadmium due to inactivation of the enzyme lysyl-oxidase. If collagen proteins are not cross-linked or bound together, you literally fall apart.

In general, enzymes containing sulfur groups or the metal zinc are inhibited or inactivated by low concentrations of cadmium. Carbonic anhydrase and alcohol dehydrogenase are two examples of zinc-containing enzymes that are inactivated by cadmium. Emphysema in industrial workers exposed to cadmium fumes seems to be due to a reduced concentration of alpha-1-antitrypsin and depressed trypsin inhibitory capacity induced by cadmium. Most of these cadmium-induced changes occur at cadmium concentrations at or below those levels found in many foods normally consumed by humans.

Welders have to be especially careful about being exposed to cadmium fumes. I was told by a visitor to our laboratory, who was a welder, of the sudden deaths of two men while welding a cadmium-coated tank. A cadmium coating looks just like a harmless zinc coating, so it is impossible to tell them apart without a chemical test. Fortunately, only 1 or 2% of dietary cadmium is absorbed. Fumes are much more dangerous with 40-50% being absorbed; hence the tragic deaths of the welders.

From a practical point of view, humans do not live in a cadmium free environment, and it seems unlikely that a cadmium free diet could be prepared. It is known that food of an embryonic nature is low in cadmium. Thus, grains, nuts, and eggs are all quite low, but still contain some cadmium. The highest amounts of cadmium are found in shellfish such as clams and oysters. Organ meats, such as kidney and liver, are also high in cadmium, especially if they come from an old animal. It seems that all organisms accumulate cadmium with age.

Cadmium and aging

In addition to cadmium in food, exposure also occurs from cadmium plated metal, nickel-cadmium batteries, some plastics, paint pigments, and the burning of fossil fuels. If it can be believed, some cooking utensils in the past were coated with cadmium in order to prevent rust. Some studies have been done involving humans. All of these studies have shown an age-related increase in the amount of cadmium in various organs, especially in the liver and kidney. A study in Sweden, where almost all of the human research has been done, reported an increase with age in liver cadmium for females and a similar increase for males, followed by a decline in cadmium after age 80 for males. Perhaps, this means that only males with low liver cadmium levels live to be more than 80. Several different studies have demonstrated an age-related increase in human kidney cadmium up to about 50 years of age followed by a substantial decline with age in some studies, but not in other studies. The question is: do people with low cadmium levels live longer?

This question led us to consider the possibility that environmental cadmium might be an important reason for biological aging. Feeding relatively low concentrations of cadmium salts reduces the life span of mice, rats and fruit flies. For example, 56 parts per million (ppm) reduced the life span of Drosophila by 58%. This is high for most food sources, but dietary kidney cortex, clams, and oysters, commonly consumed by humans, can contain this amount of cadmium. When young animals are given moderately high levels of dietary cadmium, they develop gray hair and, in general, look like old animals. The photographs of these young animals are very striking. They really look old. I have one of a calf that looks like a very old cow.

Even on normal diets, mice and Drosophila accumulate cadmium with age. In mice, just as for humans, most of the cadmium accumulation is in the kidneys and liver. We did not, however, find the decline in cadmium that has sometimes been reported for older humans. This suggests that the decline in human liver and kidney cadmium concentrations that has been observed in Sweden may be due to a selection process where people with low concentrations of cadmium simply live longer. This could be due to genetic or environmental factors. In the case of laboratory mice, the genetic background and environmental factors, such as food and water, are all the same for each mouse in any given study. This is why we did not see any decline in cadmium with aging in mice, whereas we do for humans.

Perhaps the most interesting result of the mouse study was that the cadmium content of liver increased exponentially with age. Instead of a steady linear increase with age, the amount of cadmium soared to much higher than expected levels in older mice. In aging studies you usually have two kinds of changes, either linear or exponential. If the data is plotted on the logarithmic graph paper that you can find in most college bookstores and results in a straight line increasing with time, it is called an exponential process. This is a mathematical term, which basically says that an exponential process is characterized by rapid doubling properties, which continue at an ever-increasing rate with time. The rate of doubling is called the doubling time. In contrast, a linear process displays a slow and gradual increase with time with no doubling properties.

Any exponential change with aging is of great interest since biological aging, is itself, an exponential process. The probability of death for humans doubles with every increasing 7 years of age due to the basic aging process. Any exponential process has a doubling time and for human aging it is about 7 years. The doubling time for cadmium accumulation with aging in human kidney cortex, surprisingly, is 7 years.

The exponential increase in cadmium also occurs in human livers, with the exception of the possible decrease for males after age 80, which was noted above. Cadmium, as far as I know, is the only metal ion to show this kind of behavior. In order to see if this is a general phenomenon, we looked at fruit flies maintained at different temperatures where the rate of aging is known to depend on the environmental temperature. The result was the same as it was for the mice with an exponential increase in total body cadmium with age. It is important to note that this was an exponential and not a linear increase. In this particular experiment, the median life span at 11 degrees Celsius (52 degrees F.) was 152 days, and the cadmium concentration doubling time was 64.0 days. At 30 degrees Celsius the life span was 25 days, and the cadmium doubling time was 13.5 days. At intermediate temperatures of 20 and 25 degrees the doubling times for cadmium were 35.8 and 27.5 days, respectively. So, clearly, life span and the rate of cadmium accumulation are directly related.

Cadmium and DNA destruction

I have previously described the loss of DNA from the energy producing organelles in the cellular cytoplasm, the mitochondria, which occurs with normal aging of Drosophila. Adding cadmium to the food of fruit flies greatly increases the rate of this DNA loss. The extent of the loss is dependent upon the concentration of cadmium administered. In contrast, nuclear DNA was unchanged even after long periods of cadmium feeding. Since cadmium will degrade DNA in a test tube, it is not suprising that it would destroy mitochondrial DNA, but why not the nuclear DNA as well? Apparently the site of cadmium damage is the mitochondria, and it takes the form of DNA destruction. In the nucleus this DNA damage seems to be repaired since they contain repair enzymes, whereas the mitochondria do not.

Dietary cadmium

All of this evidence suggests that cadmium accumulation could be the primary basis for aging in biological systems. Lowering cadmium in the diet is difficult, but, when we did it for fruit flies, they lived longer. We found that the lower the cadmium in the diet, the longer the life span, regardless of other dietary factors. This is unpublished data.

One of the most puzzling and interesting aspects of cadmium accumulation with aging in mice and fruit flies is that it is an exponential rather than a linear process. This must mean that something has gone wrong with the ability to excrete cadmium or that some process for avoiding absorption has broken down. Another problem is the comparison with what happens in humans. There is an exponential increase with aging in human kidney up to age 50, but then there is a dramatic decline with age after

50. The decline with age after 50 is enormous, almost a 100 fold. In the Swedish study some 80-year-old individuals had less cadmium in their kidneys than did 20 year olds. In general, cigarette smokers had higher cadmium levels. Tobacco is known to contain more than the usual amount of cadmium compared to most food sources. Nevertheless, even some 70 or 80-year-old smokers had less cadmium in their kidneys than did 24-year-old nonsmokers. Thus, tobacco use could not account for the differences. There must be some factor in the diet of people who live to be very old which prevents the age-related accumulation of cadmium. Those who died at age 90 had much lower cadmium levels than those who died at age 50.

Attempts to prevent cadmium toxicity
I wrote a grant proposal years ago dealing with this problem. It was not funded, as most basic aging research proposals are not funded. The idea was that, since bile is the principal route for elimination for cadmium, this is where the problem could be solved. Bile flow can be stimulated by various means including the administration of drugs such as spironolactone. We found that this compound increased the life span of older fruit flies by about 20%. This is unpublished data. It is also known that bile flow decreases with age in rats. Cadmium is excreted in the bile in the form of a cadmium-protein complex, but much, and perhaps all, of the cadmium is reabsorbed in the upper part of the intestines, the duodenum.

Another possible way to prevent absorption of cadmium and prevent it from being reabsorbed is to feed pectin. Pectin is a non-digestible polysaccharide which forms a negatively charged suspension called a sol. Pectin is the product in the grocery store that cooks add to some fruits for making jelly and jam. In the diet it complexes with positively charged metal ions such as cadmium to form insoluble precipitates which are then excreted. It is of some interest that pectin alone, when added to the diet, has resulted in a reduction of 15 to 18% in serum cholesterol in humans. Perhaps metal ion removal from bile by pectin is responsible for this result. The point is that the reason people with low cadmium levels live longer could be due simply to the fact that they eat something like apples which are loaded with pectin. We tried pectin in the diet of fruit flies. It gave no change in life span. Maybe this is because pectin will precipitate any metal ion, both the good and the bad.

Another approach to the problem of cadmium accumulation is antagonism by other metal ions. Zinc, which chemically is almost identical to cadmium, is known to prevent necrosis of testes if given just prior to cadmium administration. Dietary zinc is also known to reduce the negative effects of dietary cadmium. Selenium and copper also exhibit similar protective properties. The mechanism for the protective effect of selenium seems to be the formation of an insoluble salt of selenium and cadmium. Zinc and copper probably function by inducing the synthesis of the unusual protein metallothionein that subsequently binds cadmium and prevents its reaction with other molecules.

There is controversy about whether zinc or copper can actually induce the formation of metallothionein. However, there appears to be general agreement that cadmium itself is an inducer of metallothionein synthesis. It is also known that prior administration of small amounts of cadmium provides protection against much larger doses given at a later time, apparently by inducing higher levels of metallothionein. Thus, it is probable that a very low level of dietary cadmium may be of benefit to organisms periodically exposed to larger amounts of cadmium.

Since copper and zinc levels seem to be controlled by the body, they would be more desirable inducers than cadmium. At least one report, however, suggests that zinc induces metallothionein only when it is contaminated with low levels of cadmium. Therefore, it is possible that only cadmium can induce the synthesis of this unusual protein.

Certain metal ions can antagonize or prevent the toxic effects of cadmium if they are given in the diet along with the cadmium. We tried selenium, copper, iron and zinc salts. There was no improvement in life span. Even a combination of copper and zinc did nothing. A combination of iron and vitamin C also failed. Iron by itself reduced life span, as discussed above. It is of interest that the addition of ascorbic acid alone totally eliminated the 27% reduction in life span produced by iron (this is another result which we never got around to publishing). Thus, vitamin C can protect against iron toxicity but not cadmium toxicity. This is probably one reason why the combination of tea and vitamin C is so effective at prolonging life span. In animal studies, however, dietary ascorbic acid has been shown to protect against cadmium toxicity. These were not aging studies, and the doses of cadmium were higher than what we used in fruit flies. These differences need to be resolved.

In animal studies it has also been shown that exposure to low levels of cadmium, prior to exposure to a higher level, offers protection against cadmium poisoning. This is believed to be due to the induction of the synthesis of the protective protein, metallothionein. We tried giving low doses of cadmium to fruit flies one day per week. When this procedure was begun at 1 day of age, there was no significant change in their life span. However, when we tried it at 3 weeks of age, which is middle age for a fruit fly, there was an increase of 21.4% in their life span. When the experiment was repeated there was a significant 9.8% increase in life span. I, of course, was not pleased with the fact that one experiment gave an increase of 21% and another gave 10%, but, among all the metal ions we tried, cadmium was the only one to prolong life span. This was, of course, totally unexpected and not what I had in mind. Only the lesser increase in survival time of 9.8% was published in a scientific journal, but the 21.4 % value was equally valid. I just found this value to be so surprising that I wanted to be on the safe side.

I regret that we did not continue and expand this study. I did not want to be the scientist who said that a little poison is good for you, even though the dose we used was low (1.12 ppm). The fact is it seemed to be true. It is also regrettable that we did not try the same pulsing experiment using zinc or copper ions. In spite of this unusual result, however, at this time it seems that the best thing to do is to avoid cadmium exposure as much as possible.

Would it be better just to remove the cadmium? We tried. Our efforts are described in the next chapter.

Chapter 13

CHELATORS

Chelators bind to metal ions. "Chela" is Greek for claw. These molecules wrap around metal ions like a claw and prevent them from carrying out destructive biochemical reactions.

Chelators can prevent oxidative reactions from taking place. Just as an introduction, look at a jar of mayonnaise from the grocery store. The listed ingredients are: soybean oil, water, whole eggs and egg yolks, vinegar, salt, sugar, lemon juice, natural flavors (I wish that some day they would tell us what these are), and EDTA. To a chemist, this sounds like the perfect mixture for a lipid peroxidation reaction. All that natural iron and copper from the eggs and lots of polyunsaturated fat from the soybean oil would make a potent combination for producing rancid fat. Add this to the fact that these jars sit on the shelf for who knows how long without refrigeration. I use this product all the time for sandwiches and wonder why I frequently get acid indigestion after eating this product. Basically, I am too lazy to make my own fresh mayonnaise.

EDTA

The key ingredient for our story is EDTA. This is a commonly used metal ion chelator in many food products. It prevents the polyunsaturated fats and oils in the products from going rancid due to metal catalyzed peroxidation reactions. The natural copper and iron in foods can cause these foods to oxidize at greatly accelerated rates. The people who make mayonnaise and many other products add EDTA. It works to some extent. Over a period of time the product will still go rancid.

There are many natural chelators, and citric acid is one of them. It has been, and still is, used as a food preservative. The synthetic chelators, however, are much more effective. This is because they have very high binding constants for metal ions. The binding constant is a means for expressing how tightly the chelator binds to a given metal ion. EDTA has a high binding constant for lead, and for this reason it is an approved treatment for lead poisoning. Injections of this chelator have been used to remove lead from the bodies of people, mostly children, who have ingested lead.

Since EDTA is an approved medical drug, any physician can use it for any purpose. In the United States there are over 5,000 physicians who practice chelation therapy. They do it by injecting EDTA to remove calcium deposits from the cardiovascular system and for other reasons. It is amazing what they have to say about the results of this therapy. This practice continues in spite of the fact that neither the National Institutes of Health, nor any other funding agency, has ever supported even a major animal study on this very important subject.

The basis for this kind of treatment began in 1942 when Lansing reported that immersing the small aquatic animal, the rotifer, in solutions of sodium citrate greatly increased their life span. He believed that this was due to calcium removal by chelation mediated by the citrate molecules.

In 1975, Andrew Sincock published a paper in which he reported an extension of the Lansing studies. Since rotifers do not live very long, about ten days, he was able to do many experiments. The results were astonishing (Journal of Gerontology, 1975, vol. 30, pp, 289-293). Since rotifers live in the water, he simply gave them a bath in chelators for 45 seconds every other day. He used sodium tartrate and sodium citrate, which are natural chelators found in foods, and two synthetic chelators, EDTA and EGTA. Tartrate increased median and maximum life span by 43.7% and citrate by 51.7%. EDTA increased it by 49.4%, and EGTA by an almost unbelievable 75.9%. In addition, the rates of calcium accumulation with aging were reduced with all of the chelators, but not greatly. Chelators complex with almost all metal ions with calcium being just one of them. EGTA did, however, give the greatest life extension, and it had the largest calcium-binding constant of the four chelators used in the experiment.

It should be pointed out that two different laboratories have failed to find an increase in the life span of rotifers using natural and synthetic chelators. They used different strains of rotifers and different food sources. This is an issue that needs to be clarified. Since I know nothing about these tiny creatures, I am unable to offer any explanation for these conflicting reports.

Another amazing paper was published in 1967 by Wartman, Lampe, McCann and Boyle (Journal of Atherosclerosis Research, vol. 7, pp. 331-341) from a department of chemistry. They were able to demonstrate reversal of plaque in the arteries of rabbits with injections of EDTA. The paper includes two striking photographs showing the extensiveness of this reversal. The idea was that it was calcium removal, but they found no definitive evidence to support this idea. The fact is that EDTA chelates lots of other metal ions besides calcium. These metal ions include iron, copper, lead and cadmium. It is, therefore, possible that removal of other metal ions was responsible for their most interesting result.

Fruit flies and chelators
All of this evidence was enough to convince me to try chelators in the fruit fly system. Instead of giving them a bath in the chelators, we decided to add the chelators to their diet. The idea was that cadmium or other toxic metal ions were involved in the aging process, so I looked for the chelators that had the highest binding constants for these metal ions, with an emphasis on cadmium.

Since they worked on rotifers, we tried the natural chelators, citric and tartaric acids on fruit flies but found no effect. EGTA gave the greatest increase in life span in rotifers. However, when adult flies were put on a wide range of different concentrations, there was no improvement in life span with EGTA. Penicillamine is another good chelator which we tried, but it had no effect. EDTA also did not improve life span. Even two very powerful chelators, CDTA and DTPA were ineffective, and we had no success with cysteine. Thus, all of the compounds known to protect against cadmium posioning had no influence on life span, including the chelators. Needless to say, this was a great disappointment. Perhaps if the compounds had been injected they might have worked. I had to conclude that orally administered chelators alone were unable to change the life span of Drosophila.

At this point there was one other approach. The very powerful chelators we used would complex almost any metal ion. It was possible that they were creating a metal ion deficiency of some kind. In the case of cadmium toxicity, it was possible that we were removing the very metal ion that cadmium displaces which is zinc. Therefore, I decided to mix zinc with the chelators and look at the influence on life span.

When flies were fed the chelators, CDTA, DTPA or EDTA, with added equal amounts of zinc acetate beginning at 1 day of adult age, there was no improvement in life span, except for the EDTA and zinc mixture. There was a 11.6% increase in life span for low concentrations of EDTA and zinc, whereas zinc or EDTA alone had no effect.

It was puzzling that CDTA and zinc had no effect since CDTA is known to have a higher binding constant for cadmium than does EDTA. Perhaps the flies were too young. We, therefore, tried middle-aged flies with the CDTA and zinc mixture beginning at 21 days of age and removed the mixture at 41 days of age. The result was an increase in the median life span of 12.0% for the higher concentration but a 20.0% increase in life span at the lower concentration of the mixture. This indicated that the mixture had to be kept low in concentration and that the exposure time had to be limited. The survival curve was greatly improved, except for the maximum survival time (the age at which the oldest fly died), which was unchanged. This was, of course, a disappointment. If cadmium really is a basic cause of aging, the maximum survival time should have been increased as well. It was not; only the average survival time was improved.

We showed that this approach actually reduced cadmium accumulation in all age groups tested by about 50%, but it did not prevent the age related increase that occurs with normal aging in Drosophila. Thus, we have to conclude that cadmium accumulation may be a basic

reason for aging, but that chelators have limited usefulness in the control or prevention of this process. Perhaps
these kinds of experiments would work if the compounds
were injected into mice of different ages. I wish that this
had been done.

Chapter 14

COPPER

It is well known that copper is an essential nutrient and is required for the normal function of eleven different enzymes. Anemia, steely hair, central nervous system disorders, and cardiovascular and skeletal defects are among the known disorders resulting from copper deficiency in humans. On the other hand, copper is also known to be highly toxic, especially to aquatic organisms. Levels of less than 0.1 ppm copper are lethal to fat head minnows and rainbow trout. Immune response in blue gourami fish is virtually eliminated by the very low copper concentration of 0.009 ppm. It is possible that elevated copper levels are equally toxic to humans.

Free radicals
Denham Harman, who first proposed the free radical theory of aging, suggested that copper might act as a catalyst for the production of free radicals that could then initiate the degenerative changes associated with aging. Using 569 donors of various ages, he was able to show an increase with age in serum copper levels in normal human

males. This was in agreement with a previous study using both males and females.

However, other investigators were unable to confirm these results. The probable reason for this is the diverse nature of the human population, including different diets and other factors. A partial list of the factors known to affect human serum copper levels includes chronic infection, pregnancy, time of day, physical exercise, cigarette smoking, leukemia, Hodgkin's disease, iron deficiency, hyperthyroidism, ankylosing spondylitis, rheumatoid arthritis, myocardial infarction, atherosclerosis, arteriosclerosis, pellagra and psoriasis. Since it is probably impossible to find a group of humans possessing a uniform genetic and environmental make-up, inbred animal strains offer a more reliable system for these kinds of studies.

Fruit flies and mice
Since this seemed to be such an important question, we used both mice and fruit flies to examine the possible involvement of copper in the basic aging process. Total copper concentrations in Drosophila increase considerably with age by about 400%, but almost all of this increase occurs during the developmental and early adult ages. After 10 days of age there is practically no change. Dietary copper, however, had almost no influence on life span, except for very high levels of dietary copper, where there was a considerable decrease in life span. Only at levels of 60 ppm or higher was there a decrease in life span. Very few humans are exposed to this much copper in their diet. At much lower concentrations of dietary copper, using numerous salts of copper, there was no change in life span. Overall, copper did nothing to fruit flies one way or the other.

In whole organs of mice, we found that copper increases

with age in the brain by more than 45%, but decreases by 41% in the liver and 14% in the kidney. Similar results have been reported for humans. It is especially disturbing that so much copper accumulates with age in both mouse and human brains. Biochemically, brains are mostly fat, and copper is known to greatly accelerate the breakdown of fats.

Since copper is supposed to be a cadmium antagonist, we tried feeding copper at high concentrations in order to reduce the age related accumulation of cadmium. It did not work. Cadmium concentrations in liver and kidney of young, middle-aged and old mice were essentially unchanged after feeding copper gluconate for 104 days. I had to conclude that chronic consumption of copper did not prevent or reduce the normal accumulation of cadmium found in aging mice. In fact, the high concentration of copper fed to these mice reduced life span for both young and old mice. The amount of copper fed was in excess of a normal human diet. The point is that copper feeding was unable to reverse, or even slow down, the age-related cadmium accumulation.

When blood is centrifuged (spun down) to remove the red and white cells, the product is serum which is slightly yellow in color. Several laboratories have reported an increase in copper in human serum with age. I looked at mice and found the same thing, only with a much greater increase with age. I found that serum copper increased by 98% with age in mice which is much higher than that reported for humans. Since copper is a major catalyst for oxidation reactions of all kinds, this seemed like a major problem. All organisms preferentially accumulate copper in fatty tissues, and copper ions are well known to increase the rate of destructive oxidation reactions, especially for those involving fats. Based on this evidence, I thought that this must be another major cause of the basic aging process.

The copper protein, ceruloplasmin

"Free" copper in the serum may represent an important catalyst for lipid peroxidation, especially in the brain where copper concentrations are known to increase with age in both mice and humans. Free copper might also function as an inhibitor of various enzymes that have been shown to be inactivated by dilute solutions of copper ions. Among these enzymes is the antioxidant molecule, superoxide dismutase, which may be involved in the aging process as discussed in a previous chapter .

Most of the copper in serum is, however, bound to a protein, ceruloplasmin.

The function of ceruloplasmin appears to be that of a copper transporting protein, carrying copper from the liver to other tissues. It is of importance to point out that ceruloplasmin can also function as an inhibitor of lipid peroxidation. Ceruloplasmin contains 8 copper molecules per protein molecule, yielding a total copper content of 0.32%. For this reason, it is blue in color. Serum copper, not bound to ceruloplasmin, is generally regarded as "free" copper or unbound copper and is probably complexed to the protein albumin or to free amino acids.

In mice studies we found a 173% increase in serum ceruloplasmin with age. This was much greater than the 98% increase of total serum copper for the same mice. Therefore, the ratio of total serum copper to ceruloplasmin shows a considerable decrease with age in mice. This means that there is actually more "free" copper in the serum of young mice than old mice. When the data is treated in this manner, there is no significant age-related increase in the free copper content of serum with age in mice. There is actually a decrease and suggests that free copper in the blood is not a factor in the aging process. This was a surprise. Perhaps the same is true for humans.

Vitamin C
Numerous studies have shown that vitamin C inhibits dietary copper absorption. We measured copper levels in the organs of old mice that were given very large doses of 1% and 2% ascorbic acid in their drinking water for 48 days. Only the heart showed a significant decrease of 20.4% in copper and only at the 2% dose. There was no change with the 1% dose. Copper in the liver, kidney and brain remained unchanged at either dose. Dietary ascorbic acid is, therefore, unable to change the copper concentrations in the organs of older mice. It may even be toxic at high concentrations, since removing 20% of the copper in the heart is probably not a good idea.

We also looked at the effect of tea on copper toxicity. It was completely ineffective at preventing the decrease in life span in fruit flies exposed to high dietary copper. This is in contrast with its protective effect against high dietary iron concentrations.

Wilson's disease and Alzheimer's disease
Over all, it appears that dietary copper has little to do with the basic aging process. However, it is necessary to keep in mind the increase which occurs with age in the mouse and human brain. This increase is probably due to the fact that copper tends to accumulate in fatty tissues. Since the brain has a very high fat content, one would expect that it would be the primary site of copper accumulation. This seems to be true in the case of Wilson's disease where copper accumulation in the nervous tissues leads to a disease state and an early death. This illness is more

common than most people think, with about one in 400 people carrying the gene for it. An early sign is a blue ring around the iris of the eye and shaking of the hands.

People with Wilson's disease have little or no ceruloplasmin in their blood, and this may be the basis of the disease. It seems reasonable to propose that injections of ceruloplasmin might effectively treat the condition. Better yet, if one first removed the 8 molecules of copper from the ceruloplasmin by using a chelator before injecting, it might possibly remove the excess copper in the organs of these individuals.

Perhaps the same approach would work for removing the age-related copper accumulation in the mouse and human brains. It is even possible that the neurological symptoms of Alzheimer's disease are due to copper accumulation in the brain and that the administration of ceruloplasmin, with the copper removed by chelators, could be of benefit. People with Alzheimer's disease may have low levels of ceruloplasmin for one reason or another, just as people with Wilson's disease do. As far as I know, this possibility has not been investigated.

It seems that the question of copper and aging remains unanswered in spite of our efforts and those of others to answer this question.

Chapter 15

ALUMINUM

Aluminum is one of the most abundant elements. It is remarkable that the aluminum concentrations in biological tissues are so low when considering that aluminum is the third most abundant element in the earth's crust. In its metallic form, it is widely used in many products including aircraft, machine parts, and soda and beer cans.

Most aluminum in the earth's crust takes the form of insoluble salts such as aluminum oxide. This material is used in the grinding wheels of shop tools since it is harder than steel. These insoluble salts are also used to make many kinds of water-soluble aluminum compounds. Antacids and baking powders are two of the most familiar uses of soluble aluminum salts. In the past it was generally believed that these aluminum salts were not harmful and that aluminum did not accumulate with age in humans.

Aluminum, nerve disease, and osteoporosis

These general beliefs for the non-toxicity of aluminum came into serious question in 1973, when it was shown that aluminum intoxication in cats gave symptoms similar to the memory loss occurring in Alzheimer's disease.

In 1982, it was reported that high dietary aluminum is associated with Lou Gehrig's disease and Parkinson's disease on the island of Guam, where the dietary intake of aluminum is unusually high. Changes in the water supply for this island have recently been made in an attempt to correct this problem. It has also been reported that elevated aluminum levels are associated with soft and brittle bones in people who consume a large amount of aluminum-based antacids for several years. Alzheimer's disease is more prevalent in areas, such as Wales, with elevated concentrations of aluminum in the drinking water. People with kidney failure who undergo dialysis treatment sometimes suffer from dialysis dementia, the cause of which has been found to be aluminum intoxication resulting from their inability to excrete aluminum in the urine. Contamination by aluminum in some cases of the equipment used for dialysis has also caused dementia. However, it should be noted that dialysis patients almost never develop Alzheimer's disease.

Aluminum and aging

It is interesting that these two different disease conditions of dementia and osteomalacia should both be induced by aluminum salts. Perhaps the basic aging process is responsible for both of these conditions and that accelerated aging induced by aluminum is the reason. Aluminum is known to accumulate with age in dogs, and two different studies have shown that it accumulates with age in normal human brains. We found that it accumulates with age in fruit flies in a linear manner by about 50%.

In view of the great interest in this subject, we examined the aluminum content of the organs of mice of different ages. Total aluminum concentrations increased with age in the liver and kidneys of mice, but remained unchanged in the brain and heart, and decreased with age in bone and lung. While the amount of aluminum in bone decreased with age, it contained the greatest amount of aluminum with average values of 15 ppm (parts per million). Much higher values have been reported for human skin and lung. A value of 94 ppm has been reported for human lung, whereas our highest concentration for the mouse lung was less than 6 ppm. High values have also been reported for aluminum in the lungs of old dogs, while only low concentrations have been found in young dogs. These researchers proposed that the source of the aluminum was dust in the atmosphere. Atmospheric aluminum exposure in humans seems to be due primarily to dust and dirt. It is, after all, the third most abundant element on the surface of the earth. Apparently, atmospheric aluminum was low in our mouse colony. We also found that aluminum in the lung actually decreased with age in mice.

A more recent report of aluminum in the lung of humans versus age gives an average of 56 ppm with an increase with age. This is still very high but suggests that the human exposure to atmospheric aluminum may be declining. The people who make those shop tools with aluminum oxide grinding wheels, for example, now recommend that you wear a mask when using these machines.

The brain is another organ where we had expected to find an increase in aluminum with age. There was no change with age, which was in agreement with a report of no change in the aluminum content of human spinal fluid with age. However, two different laboratories have demonstrated an increase with age in human brain concentrations of aluminum, with the major part of the increase occurring after 70 years of age. In contrast, another laboratory found no increase in aluminum in the brains of normal humans with age.

We were given 5 very old mice by the National Institutes of Aging. Only 3 survived shipment. They were 1186 days of age, which is very old for a mouse. Both their scientists and ours had expected that these very old mice would have elevated levels of aluminum in their brains. They did not. They did, however, have the highest aluminum levels in their livers that we had ever found in mice. Perhaps the problem of aluminum accumulation with age is in the liver and not in the brain. Therefore, we have to conclude that accumulation of aluminum in the brain with age does not appear to be a general phenomenon. This is an inconsistency that needs to be resolved.

The liver and kidney were the only organs in mice to give an increase in aluminum concentrations with age. However, the increases were small. Aluminum in the heart did not change with age, and aluminum in the bone and lung actually declined with increasing age.

The major means for excreting aluminum is through the urine. Kidney function is known to decrease with age. We, therefore, tried to change the aluminum content of the various organs by altering kidney function. The result was that all organs had an increase in aluminum, but the differences were not statistically significant. Increased aluminum concentrations in the liver and kidney are, therefore, probably not due to age-related decreased kidney function, at least in this mouse system.

It should be pointed out that aluminum might be an essential nutrient. It is known that aluminum ions are required for the enzyme, adenylate cyclase. This is an important enzyme required for normal metabolic function. Aluminum may also have a nutritional role based on the cholesterol lowering effects of aluminum-polysaccharide complexes. It could be that there is an optimum intake of dietary aluminum for humans, but we just do not know what it is. We know what it is for fruit flies (2.7ppm).

When we doubled the aluminum dietary intake for mice beginning at 604 days of age, which is old for a mouse, there was a small reduction in life span of 6.7%. This was similar to the result we found for fruit flies. Therefore, it is clear that excessive dietary intake of aluminum salts can increase the rate of aging and moderately reduce life span. It has also been reported that the natural chelator, citric acid, greatly increases the absorption of dietary aluminum by 50 fold. Alzheimer's disease may be more prevalent in people who drink juices containing large amounts of citric acid such as orange and grapefruit juice. It is probably not a good idea to take an aluminum based antacid tablet with a glass of orange juice. In fact, it is probably not a good idea to consume any aluminum-based medicine if there is an alternative.

We can conclude that aluminum exposure increases the rate of aging and that it is probably involved in disease states, especially several forms of dementia. It has been known from as early as 1897 that aluminum affects brain function, yet very little has been done to address this problem. There is no doubt, at this late date, that high intakes of aluminum salts can produce osteomalacia (soft bones) and dementia in humans. Less than 1% of all humans suffer from Alzheimer's disease; however, more than 20% of people over the age of 80 have it. Clearly this is an age-related disease, as are cancer and heart disease. Children and young people develop cancer and heart disease, but not Alzheimer's disease. The study of Alzheimer's disease offers a real opportunity to understand the basic aging process. If aluminum really is the reason for Alzheimer's disease, there may be ways to prevent or reverse it.

A theoretical proposal

Although based on good evidence, what follows is pure speculation. As noted in the iron chapter, there are iron-transporting proteins in the blood. One of these is ferritin, and another is transferrin. Transferrin delivers iron to the cells in the body. It also has been shown to be a powerful and natural chelator of aluminum. Normally, transferrin carries iron, and it has many sites for complexing iron. If these sites are already occupied by iron, they cannot complex aluminum ions. At a high concentration of total body iron as occurs in most older people, most of these sites are occupied. Conversely, a low body iron content, which is the case for most young people, would leave many of these sites open to other metal ions such as aluminum. It is known that aluminum in the blood is bound almost exclusively to the plasma glycoprotein, transferrin. The theory that has been proposed is that aluminum ions not bound to transferrin will be able to enter the brain.

It is known that people with Alzheimer's disease do have transferrin molecules which bind much less aluminum than transferrin from normal people. Thus, the assumption is that more free aluminum is available to enter the brain. It is not clear why their transferrin is different, but it does contain twice the amount of iron as transferrin from normal people. Alzheimer patients have transferrin molecules that contain even as much or more iron than people who suffer from the iron overload disease of hemochromatosis, yet their ferritin levels, which indicate total body iron content, are normal. Reducing the iron content of their transferrin molecules might produce some interesting results. It could be as simple as donating blood. Perhaps such a procedure could prevent, or even cure, the disease. It is not unreasonable to propose that Alzheimer's is an unusual form of iron overload disease.

As far as I know, no one has done any research from this point of view.

Dietary aluminum

It is probably prudent to avoid excessive dietary intake of aluminum salts such as those found in many antacids and most baking powders. Phosphate-based baking powders that contain no aluminum can be purchased. I remember a woman who came to my office and wanted to tell me the story of her husband who was previously employed as a restaurant manager. He came down with Alzheimer's at the relatively young age of 50. It all started when he called her from a phone booth saying that he was confused and did not know how to get home. She related to me that he had for years consumed 3 rolls per day of a popular aluminum based antacid. She was convinced that this high intake of aluminum had given her husband Alzheimer's. This was, of course, a tragedy, but it is difficult to attribute it solely to the aluminum in his antacid pills.

Another possible source of high dietary intake of aluminum is from aluminum cans used for soda and beer storage. We tested various beverages from these containers and found essentially no free alumium in the products. There was actually more aluminum in beer and soda from glass bottles, but at very low concentrations.

Aluminum in the air

Overall, it can be concluded that moderately elevated dietary aluminum increases the rate of aging in mice and fruit flies. There is also an accumulation of aluminum in mouse organs with age but mostly in the liver, and to a lesser extent in the kidney. There is no increase with age in mice in either the bones or the brain. The aluminum content of fruit flies increases by 50% with age. Unlike humans, these mice and fruit flies were kept in a controlled environment. Since most humans are exposed to a lot of dust in their environment, this probably explains why humans accumulate aluminum in the brain and lungs with age, and experimental mice do not.

The average human diet contains 20 milligrams of aluminum per day, but only 5% of that is absorbed. That would amount to only 1 mg of aluminum per day, less than the weight of a single fruit fly. It seems almost unbelievable that this small amount of aluminum could account for the fact that humans accumulate aluminum in the brain with age. There must be another source. Antiperspirants containing aluminum chloride are widely used by humans, and human skin does contain a high amount of aluminum, but it seems doubtful that this aluminum is absorbed into the body. We cannot, however, discount this possibility. Further examination is needed.

The amount of aluminum in some human lungs is very high and has been shown to increase with age in an exponential manner. Again, it is very important that this is an exponential, rather than a linear, process since aging is an exponential process. One 70 year old, for example, had over 300 ppm of aluminum in his lungs. A food source containing this amount of aluminum would decrease the average and maximum life span of fruit flies by as much as 20%. It seems highly probable that the main site for entry of excess aluminum ions in humans is the lung as a result of breathing dust.

There is no doubt, based on numerous research reports, that excess dietary aluminum causes dementia and extensive calcium loss from the bones in animals and humans. Aluminum, like copper and iron, is an essential nutrient, but excess exposure, either in the diet or from the air, can increase the rate of aging or produce some disease states.

It is possible that Alzheimer's disease is one of these disease states. Our suvival studies with Drosophila suggest that 2.7 ppm should be the upper limit for aluminum in the diet. People exposed to high levels of dust should wear a mask.

Chapter 16

CHROMIUM

Chromium is another essential trace element in the diet. It is the sixth most abundant element in the earth's crust, and, as a result of its abundance, deficiency is seldom seen. In its metallic form, it is used to plate steel and other metals in order to improve their appearance and to inhibit rust. In its nonmetallic, or ionic, form it can exist in two different states, one with a plus 3 charge and another with a plus 6 charge. Chemists call this the valence of a metal ion. Chromium with a plus 6 charge is highly mutagenic (causes DNA damage) in bacteria and, therefore, is most surely mutagenic in higher organisms. It has been shown to cause cancer in laboratory rodents. Chromium with a plus 3 charge, however, is not mutagenic and does not cause cancer. Basically, chromium with the plus 3 charge is good and chromium with the plus 6 charge is bad. The easiest way to convert plus 6 chromium ions to plus 3 metal ions is to simply mix in a little vitamin C. This is another reason why a little extra vitamin C in the diet is probably a good idea.

Chromium poisoning

While chromium is an essential component of the diet, excessive chromium exposure can be quite toxic. Chromite miners, chemical workers, and anodizing and leather tanning industrial workers have experienced chromium poisoning. The symptoms include skin ulceration, allergic dermatitis, inflammation and perforation of the nasal septum and larynx, repiratory tract cancer and various other systemic disorders. Ascorbic acid has been used to treat and prevent these illnesses, which means that chromium in the plus 6 valence state is probably the source of the problems. The chelators, EDTA and DTPA, have also successfully been used to treat chromium poisoning by removing it from the body.

Chromium in heart disease and cancer

Since chromium has been implicated in both atherosclerosis and cancer, the two leading causes of death in the elderly, it is of considerable interest from the point of view of aging research. Chromium appears to be an unusual dietary trace element in that it is implicated positively in the prevention of cardiovascular disease but negatively in the development of cancer. This apparent inconsistency can be explained based on the two forms of chromium. The trivalent state has a positive effect, whereas the hexavalent state has a negative effect.

High mortality from coronary heart disease has been reported for the eastern area of Finland where the concentrations of chromium in drinking water were found to be low. In the western area of the country, chromium levels were higher and coronary heart disease less prevalent.

In another study, low serum levels of chromium were found to be associated with atherosclerotic disease. In rats on chromium deficient diets, aortic plaques and elevated glucose and serum cholesterol have been observed. The proposed reason for this is that chromium is the active component of glucose tolerant factor (GTF). This factor, which is found in brewer's yeast, lowers the amount of insulin needed to remove sugar from the blood. Diabetic conditions appear to be improved by the addition of this factor to the diet. It seems to lower the amount of insulin required to control diabetes. The assumption is that, in normal people, GTF will help to maintain normal blood sugar levels and thereby decrease the incidence of cardiovascular disease, since high blood sugar promotes cardiovascular disease. This could be the reason that dietary chromium lowers the incidence of coronary heart disease.

The negative side of this story is that epidemiological evidence suggests the involvement of chromium in lung cancer among workers in chromium refineries and related industries. Also, lifetime feeding of 5 ppm of hexavalent chromium (the plus 6 chromium ion) results in the formation of malignant tumors in mice. As was already noted, hexavalent chromium is both mutagenic and carcinogenic. Therefore, this mouse result was to be expected. The presence of pulmonary, or lung cancer, as opposed to other forms of cancer, suggests that the industrial workers were exposed to dust containing chromium which was deposited in their lungs. If this chromium was in the hexavalent form, and it probably was, then they were exposed to a carcinogen, and the terrible result was pulmonary cancer. In the diet, perhaps some of this hexavalent (plus 6 form of chromium) chromium would have been converted to the trivalent form (plus 3 chromium ion) by vitamin C, and this is why they had pulmonary rather than other forms of cancer. Just as is the case for aluminum, exposure of humans to dust containing metal ions of chromium can lead to health problems.

Chromium and aging

A major reason for considering chromium significant in the aging process is the fact that it declines dramatically with age in a variety of human tissues. Most of this decline occurs at a young age (20 years or less), and it is not clear whether or not the decline continues into old age. In fact, the human lung shows a considerable increase with age. This is probably due to human exposure to dust. There is, however, little or no change in the chromium levels with age in human blood.

In view of the problems associated with looking at an uncontrolled human population, we decided to look at the changes in chromium associated with age in fruit flies and mice. To our surprise, we found no significant change in total chromium in adult fruit flies with age. Chromium levels were higher in the developmental stages where they were more than twice the amount found in the adult stage.

We also looked at the chromium concentratons in our colony of mice. The chromium concentration in liver increased by 25.2% from 45 to 886 days of age. Brain chromium decreased by 19.9%, but the decline was not statistically significant due to scatter in the data. Heart and kidney showed absolutely no change at all with age. Since we used an unusually large number of mice for this study, there can be little doubt about the results. Of course, these results are in total disagreement with the results reported for humans which showed a very large decline in chromium concentrations with age. Since diet and environment are not controlled in human studies, these studies must be regarded with a degree of caution.

In agreement, however, with our results for fruit flies and mice, are the two reports of no change in chromium in the blood of normal humans with age.

There are two published reports demonstrating that prolonged feeding of chromium in the drinking water results in an increase in life span of rats. It is of interest and considerable surprise that the hexavalent form (the plus 6 chromium ion) of chromium was used in both of these studies. This is the mutagenic and carcinogenic form of chromium. Both of these studies were aimed at looking at the toxic effects of chromium. It was, therefore, appropriate that the hexavalent rather than the trivalent form of chromium was used. I am sure that the investigators were surprised at their results. Perhaps the reason for such an unexpected favorable result was that rats produce vitamin C, which could have converted the low levels of hexavalent chromium to the trivalent form, which is not mutagenic or carcinogenic.

Whatever the reason, I was encouraged to try feeding the trivalent form of chromium to fruit flies in order to see if there would be a change in life span. There was none. Lifetime feeding of trivalent chromium at a 100-fold range of different concentrations failed to produce any significant change in life span. Perhaps the only positive result of this experiment was that there was no significant decrease in life span at any of the concentrations tested. Chromium, therefore, at least in the trivalent form, is relatively nontoxic when compared to other metal ions such as cadmium and iron.

In conclusion, our results offer little or no support for the possibility of chromium deficiency being involved in the basic aging process. The only organ to show a significant change with aging was mouse liver, and there we found an increase rather than a decrease with age. In fruit flies, chromium content did not decline with age and supplementing their diet with a wide range of chromium concentrations had no influence on life span.

The case for chromium deficiency being involved in human cardiovascular disease is, however, compelling. Certainly, mouse or rat survival studies should be done with trivalent chromium and the chromium containing glucose tolerant factor (GTF). This factor is also involved in diabetes and probably, as a consequence, in cardiovascular disease since diabetes leads to cardiovascular problems. These possibilities represent a great opportunity for studying the involvement of chromium in disease states and aging.

Food sources of chromium
The cardiovascular disease research results suggest that it might be desirable to add a little chromium to the diet. Fruits and vegetables are very low in chromium (about 0.01 ppm) which is surprising since there is so much chromium in the soil. Meats and eggs are ten times higher (0.1 ppm), but the greatest amount of chromium is in spices. Black pepper is 3.7 ppm chromium, and thyme is the highest of all at 10 ppm. The best source of chromium containing GTF (glucose tolerant factor), however, is brewer's yeast.

Chapter 17

LEAD

Among all the metal ions, lead has the worst reputation, but this is only recently. Humans have used metallic lead and lead salts for thousands of years. The Romans used it for drinking vessels and to line their famous water aqueduct systems. Some of these aqueducts are still in use today, hopefully minus the lead linings. Incidentally, the average life span during the height of the Roman Empire was only 25 years and more that one historian has suggested that the collapse of that civilization was due to lead poisoning. Until very recently, lead pipes were used to bring water into many homes. A large number of these pipes are still in use today. Lead based solder was, until recently, widely used in pumbing and still is in electronics. Most old paints and dyes contained lead, and ceramic coatings contained a great deal of lead. Some still do.

Major sources of lead contamination
In 1975, 160,000 metric tons of lead were emitted into the atmosphere in the United States from the burning of gasoline containing tetraethyl lead. This was an anti-knock compound that made gasoline engines run more

smoothly. Surprisingly, however, the Envonmental Protection Agency in 1977 stated: " Probably the largest proportion of dietary lead is derived from food processing (e.g. from solder in the seams of cans), and some is also derived from lead in the air and the soil." Today there is no lead in the gasoline of most countries, and many, but not all, food processors package their products in die-punched cans which contain little lead.

The legacy of lead poisoning, however, still persists. Lead salts have a sweet taste similar to sugar. In the past, people in Europe, especially in England and France, consumed chocolates coated with colorful and tasty lead salts. They either died or suffered from lead poisoning. Today children eat lead paint for the same reason. It has a sweet taste. Many older houses contain lead based paint that has not been covered or removed. Much of the neurological damage to these children has been shown to be irreversible. There are still many sources of unnecessary lead exposure that must be eliminated. Fortunately, this is actually being done for the first time in human history.

Lead and aging
Relatively low-level exposure to lead has been implicated in many disease states associated with the aging process. High blood pressure, abnormal electrical and mechanical activity of the heart, atherosclerosis, decreased antibody formation, senile cataracts, osteoporosis and behavior impairment have all been ascribed to elevated dietary lead. All of this suggests that dietary lead may be involved in the development of senescence.

Human studies have shown that lead accumulates with age in various organs. It has also been shown that adding lead at 5 ppm (parts per million) to the drinking water reduces the life span of both mice and rats.

We looked at the lead accumulation of fruit flies on a normal diet and found that it increased by 173% with age. This is a very large increase considering that they were not exposed to any source of lead other than food and air. I do not know where it came from. Since most of the lead in mammals accumulates in the bones, we examined the lead content of the bones of mice and found that it increased by 83% with age. Both of these results are in agreement with the reported increases of lead in human organs with age.

In humans the total body lead content increases with age by an incredible 10,000% between the ages of 0-9 years and 60-69 years (Barry and Mossman, Brit. J. Indust. Med., vol. 27, pp. 339-351, 1970). This is a large increase when compared to laboratory animals. The total amount of lead in both fruit flies and mice of old age was greatly lower than the values for humans. Tissues from humans contain an incredible 100 times more lead than these organisms. Humans are being exposed to a very large amount of lead, and some of the sources remain unknown.

A reputable laboratory has demonstrated that lifetime feeding of 5 ppm of lead in the drinking water reduces the life span of mice by 17% and that of rats by 34%. We tried to do the same thing with Drosophila and found no change in life span at this concentration. We also tried a wide range of other concentrations differing by a thousand fold. There was no change in life span until we reached the very high concentration of 2070 ppm, where the life span was reduced by 19.3%. Since there was no change in life span at very low concentrations, it is unlikely that lead is an essential trace element with some, as yet unknown, favorable function in terms of the aging process. However, it should be noted that the very high tolerance of fruit flies for dietary lead is almost unbelievable.

Lead and cardiovascular disease

While most of the lead accumulation with age in humans is in the bones, a large amount of lead accumulation with age has been found in the human aorta. The aorta is the major blood vessel leading from the heart where plaque formation is common in people with cardiovascular disease. Plaque is a hard substance with an appearance almost like a tumor, composed of living cells, calcium, iron, fat and cholesterol. Feeding high levels of lead has been shown to produce plaques in the aortas of animals. Plaques reduce blood flow and are the sites where blood clots can form leading to heart attacks or strokes. Since lead can produce plaques in the aorta, it is not unreasonable to propose that it also produces plaques throughout the cardiovascular system. As noted above, lead also increases blood pressure. Plaque formation and elevated blood pressure are the two most serious risks for heart attack and stroke, and they both are induced by high intakes of lead.

It is of some interest that in the United States the major causes of death in 1900 were pneumonia and influenza, followed by tuberculosis and diarrhea. Diseases of the heart and stroke were only number 4 and 5. Tetraethyl lead was introduced into gasoline in 1923. By 1940, diseases of the heart had become the number one cause of death, followed by cancer and stroke. In 1960 diseases of the heart increased dramatically, again followed by cancer and stroke as numbers 2 and 3. In 1970 the results were similar. In 1980, there was a slight decine in heart disease, a considerable decline in stroke with an increase in cancer, but they still remained in the top three.

It should be noted that the average life expectancy in 1900 in the United States was only 42 years of age. Why did the death rate from diseases of the heart increase from

137 per 100,000 in 1900 to 292 per 100,000 from 1900 to 1940 and reach a high value of 369 per 100,000 in 1960? One possibility is the large increase in lead exposure during this time period. Lead produces cardiovascular disease in animals, and it also reduces immune response. This reduction of immune response could be related to the very high increases in cancer observed during this period of time. Is lead a cause of cardiovascular disease or cancer? The answer is clouded by other events which happened during this time period.

The introduction of sulfa antibiotic drugs to combat infections allowed people to live longer. Prohibition of alcohol and the influenza epidemic of 1919-1920 have been suggested to be the basis of the sudden increase in cardiovascular disease during this time period. One theory of cardiovascular disease is that viruses cause it. Moderate alcohol consumption is also known to decrease the incidence of this disease. Perhaps it was simply due to an increase in life span that resulted in a change from infectious to degenerative diseases as the leading causes of death. The purpose of this book is to deal with the basic aging process, not disease states, but the involvement of cardiovascular disease in the human aging process cannot be ignored. One third of all the deaths in the United States are due to cardiovascular disease. Almost all of these deaths occur in people over the age of 40. In spite of years of research, the real cause remains unknown. There are drug measures and surgery to control the disease, but basically it is still here in full force. Our result with fruit flies indicates that lead has little to do with the basic aging process in fruit flies. Nevertheless, it could be involved in the disease states of cardiovascular disease and cancer and even aging in humans.

There are two important papers related to this subject.

One is by Revis, Zinsmeister and Bull, in the Proceedings of the National Academy of Sciences, USA, vol. 78, pp. 6494-6498, 1981. Another is a paper by Settle and Patterson in Science, vol. 207, pp. 1167-1176, 1980. In my opinion, this latter paper is the best paper ever written on the subject of lead and health. A technical background is not necessary to appreciate its quality.

The paper on lead and atherosclerosis by Revis, Zinsmeister and Bull is consistent with the observation of lead accumulation in human aorta with age. Feeding pigeons 0.8 ppm of lead in their drinking water for 6 months resulted in a large increase in the number of atherosclerotic plaques in their aortas. Since pigeons can live for 15 or more years, this was not a long time period compared to their total life span. This amount of lead exposure is probably less than the exposure many humans experience, according to the authors of this paper. The amount of lead in a lead soldered can of tuna fish is 1.4 ppm, and for anchovies in a lead soldered can it is 4.2 ppm. Die-punched cans are much lower, with 0.007 ppm lead content for tuna fish.

Lead and the environment

I recently watched a Public Broadcasting Program on an English expedition to the Arctic more than a century ago. Their source of food was canned goods, which at the time was considered a marvelous new invention. Almost everyone on the expedition experienced dementia and eventually died from lead poisoning. They had no idea what was going on. Even today the greatest source of lead intoxication is from canned foods, according to the Environmental Protection Agency as noted above.

The Settle and Patterson article is disturbing. They point

out that the lead content of Peruvian Indians living 1800 years ago was 500 times less than the lead content of humans today. They also make the statement that "an unrecognized form of lead poisoning may be affecting most Americans and a major portion of the world's population." They also state that regulatory agencies have failed to regulate lead levels in the human population with serious consequences to human health.

Lead consumption throughout the world has never been higher. Every year 3 million tons of lead are produced worldwide. Even at the height of the Roman Empire lead consumption was 4 kilograms per person, versus 6 kilograms per person per year today in the United States.

The average lead content of human blood is 0.2 ppm. At a concentration of 0.4 to 0.6 ppm serious neurological symptoms of lead poisoning appear in humans. This leaves very little margin of safety from neurological lead poisoning. The problems of high blood pressure and atherosclerosis are also of considerable concern and apparently occur at even lower levels of lead exposure.

It is obvious that fruit flies are not a good model system for studying lead intoxication and aging. It is not clear why they are so resistant to dietary lead. Perhaps, it is because they have no bones and, therefore, do not accumulate lead as humans and animals do. Over 90% of the lead in humans is stored in the bones with men having about 50% more lead in their bones than women. The amount of lead in human bones is almost 10,000% higher than what we found for laboratory mice. This must mean that humans, and especially men, are exposed to high levels of environmental lead.

Atmospheric lead is also another source of exposure. This lead is deposited in the soil and in the polar ice caps. The amount of lead in the upper 10 meters of the Greenland ice cap was 10 tons in prehistoric times. Today, the top 10 meters of that ice cap contains 4,000 tons of industrial lead. The soil of the Northern Hemisphere used to contain 300 tons of lead. Today it contains 40,000 tons. Trees now contain 10 times the amount of lead as they did in the past. Humans now absorb into the blood 100 times the amount of lead they did in prehistoric times.

There is evidence for lead being involved in heart disease, stroke, high blood pressure, reduced immune function and osteoporosis. In terms of human history, these are all relatively new diseases, and it is possible that lead exposure is responsible. It has been suggested that senile dementia is also related to lead poisoning, based on the fact that bone loss usually occurs with aging. That, in turn, would release lead into the blood and brains of older people. Since 90% of the lead in humans is located in the bones, this makes sense to me.

In conclusion, the reduced life span in laboratory mice and rats fed excess dietary lead suggests that elevated lead intake is involved in the aging process in mammals. It is not known what concentration of lead will do the same thing in humans. It is clear, however, that humans accumulate very large amounts of lead in their bones as they grow older, 100 times more than laboratory animals. It would be prudent to avoid lead exposure as much as possible since it has no known beneficial biochemical function and seems to cause a variety of health problems.

Chapter 18

BORON

Another element which may be involved in the aging process is boron. The involvement may be in both a positive and negative way. Boron is not a metal. For hundreds of years boron compounds in the form of sodium borate and boric acid have been used as food preservatives, eye washes, cleaning agents, fungicides and insecticides. Boron kills bacteria, and this is the basis of its use as a food preservative and for eyewashes. It gives a cleaner laundry probably for the same reason. You can use it to remove mold in the bathroom, and it will kill almost all insects at relatively high concentrations. It has been used to control cockroaches, and a mixture of borax and sugar dissolved in water can be used to eliminate ants from the home. The ants carry it to their nest where it kills the queen and the eggs in the nest. It also slowly kills the adult ants.

In the form of boric acid, boron is also used in the cooling water surrounding nuclear reactors in order to protect us from the radioactive neutrons produced by the nuclear reaction. Neutrons are very dangerous particles and the use of boron is the best way of capturing them. Boric acid, like most acids, is corrosive. I just read that a nuclear reactor had to be shut down because the steel containment vessel was punctured by the action of boric acid. The salt of boric acid is sodium borate. This is the form of boron with which we are most familiar and can be purchased in almost any grocery store for laundry and other household uses. The common name for this product is borax.

Boron in agriculture

It has been known for many years that boron is an essential nutrient for all plants. In areas with high rainfall the boron content of the soil is generally low due to leaching of boron from the soil. In these parts of the world, boron is usually added to fertilizer preparations. In more arid climates the boron content of the soil can be quite high and may reach concentrations where plants die from the excess boron. This is especially true in areas where irrigation water is used in agriculture. Boron has successfully been used as a herbicide to control weeds based on its toxicity for plants at high concentrations.

Boron and osteoporosis

Until 1987, it was believed that animals and humans had no requirement for dietary boron. Then it was reported that boron deficiency might be the cause of the bone wasting disease of osteoporosis (Nielsen, Hunt, Mullen and Hunt, FASEB J., vol. 1, pp. 394-397, 1987). It was also reported by these scientists that chickens developed abnormally soft bones in the absence of dietary boron. It had also previously been reported that rats had bones that broke more easily on boron deficient diets. This was an important discovery and suggested that osteoporosis, which occurs in aging animals and humans, could be due to boron deficiency.

Most dietary boron comes from fruits and vegetables.

Vegetarians seldom develop osteoporosis and Eskimos, who eat very little in the way of plant foods, have a very high incidence of osteoporosis. All of this seemed to suggest that boron is some how involved in the aging process.

Boron is used in a wide range of consumer products. It is not clear whether or not boron in these forms represents a health hazard. One long-term study with boron resulted in coarse hair coats, scaly tails and hunched position when rats were given a high dose of boron in their food for one year. It is of interest that these symptoms are similar to the signs of normal aging in rats. In another study where rats were given a high concentration of boron in their drinking water for 70 days, testes, spleen and bone weights were greatly reduced. Yet, it has been shown that boron deficiency, not its excess, may be a cause of osteoporosis.

One possible explanation for this apparent inconsistency is that animals may respond to boron in a manner similar to plants. Boron is essential for plants, but it is also a source of toxicity. The optimum range for plant growth is narrow and variable. Lemon trees are very sensitive to boron, but beets are not. The mechanism for the effect of boron deficiency and toxicity in plants has not been elucidated. In general, plant scientists describe boron toxicity as accelerated plant senescence.

The reports are widespread and apply to numerous species. For example, in pear trees, 0.14 ppm boron in the soil results in normal tree growth, whereas 0.56 ppm boron leads to dieback of the plant shoots. Compared to most other plant nutrients, this is a very narrow range of benefit versus toxicity. Crop yields can be reduced by more than 50% in boron deficent soils, but they also can be reduced by an equal amount in soils with excess boron. This, of course, represents an issue of great economic importance to farmers and has, therefore, received considerable attention.

The leaves of deciduous trees accumulate boron during the growing season, reaching a peak in the fall. It has been proposed that shedding of leaves of these trees in the autumn may represent a means for boron excretion in plants. The mechanism of leaf senescence as a result of boron accumulation remains unknown.

There is a report that increasing the boron concentration in plant growth medium results in the formation of atypical and enlarged mitochondria. The changes in mitochondrial appearance are similar to those seen in normal aging of insects and mammals.

The extent of long-term boron toxicity in economically important plants appears to be well documented but not understood. In animals and man, much less is known about the long-term exposure to boron. The increased rate of aging observed with excess boron in plants may also occur in other organisms including humans.

Fruit flies
We looked at fruit flies and mice. We found that very low boron diets led to a faster rate of aging in Drosophila, but that high boron diets also greatly accelerated the rate of aging. This was all very similar to what had been previously reported for plants.

We found that high boron concentrations in the diet of Drosophila decreased life span by as much as 69%, whereas lower, but not very much lower, boron concentrations added to the diet increased average and maximum life span by 9.5%.

Dietary boron

The boron content of foods varies enormously. Generally the levels are quite low. Very high values have been reported for caviar, where boron has been used as a preservative. Most meats and fish have a low content, but there are very large variations in the values reported. Lettuce has almost no boron, but other vegetables and fruits have much more. Grains are generally very low but extremely variable. This is probably due to different soil contents of boron.

Probably the best article published on the boron content of foods is by the French professor J. Ploquin, Societe Scientifique D'Hygiene Alimentaire, vol. 55, pp. 70-113, 1967. He reported that an average French diet contained 49 milligrams of boron per week with 20 milligrams coming from vegetables and 16mg from fruits. Please remember that there are 1,000 milligrams to a gram and 454 grams to a pound. This amounts to 73% of dietary boron coming from fruits and vegetables. Potatoes were also high, contributing 3mg. Milk, cheese, meat, fish, eggs, oil and butter were all very low, and sugar contained no boron at all. Total food intake for a week amounted to 14,990 grams with a total boron intake of 49 milligrams. The ratio is 0.000,003, which means that the total diet contained 3 ppm boron on a wet weight basis. Wet weight means that the food was not dried out before analysis.

In order to compare this with our fruit fly results, we had to convert to a wet weight basis since we dried the food out before analysis. Our maximum life span for Drosophila was achieved at a concentration of 11.3 ppm boron on a wet weight basis. This means that the French boron intake is 4 times lower than that of the fruit fly diet that gave the maximum increase in life span. This is not a great difference unless you consider the very narrow range of boron benefit versus toxicity for plants and whether or not this extends to humans. In plants it takes very small increases in the boron content of the soil to produce considerable changes in growth and crop yield. A similar small addition of boron to the human diet could result in a great change in overall general health and life span.

The real questions are: what is the human requirement for boron and why in the world is it needed? There are no known enzymes that require boron and no known biochemical reactions requiring boron. The human osteoporosis studies noted above used 3 mg of boron per day as a supplement, and yet the French results indicate that the average human intake of boron is only 7 mg per day. It is striking that such a small addition of boron to the diet could decrease calcium loss from the body and favorably change hormonal levels in the individuals who participated in this study. Boron must be a powerful compound, and even a very moderate deficiency probably leads to serious health consequences, such as osteoporosis and aging.

Boron deficiency
It is not unreasonable to believe or even to propose that boron deficiency is the cause of osteoporosis, and yet there has been very little research on this subject. The question of boron's involvement in the aging process has been essentially ignored. Perhaps because only until recently has the possibility of boron being an essential nutrient in animals and humans been considered. Boron deficiency in humans is probably widespread. Since boron can be toxic and can increase the rate of aging in both plants and animals, dietary supplementation at this point could be hazardous. In general, most soils are boron deficient, so farmers use large quantities of it in the form of fertilizers to increase crop yields. Even some compost used by

organic farmers is very high in boron and sometimes too high. Some kinds of compost can cause boron toxicity in plants. Since farmers are using boron for crop production, the safest approach to dealing with possible boron deficiency is to eat more vegetables, fruits and potatoes.

The conclusion is that we know almost nothing about boron and its relationship to human health and the aging process. The only thing which can be said with confidence is that fruit flies consuming a diet containing 11.3 ppm boron on a wet weight basis will live 10% longer. At higher concentrations life span will be reduced, and the same is true at lower concentrations. At a conference on boron and health, I was asked what should be the human intake of boron. My answer was that I did not know, and I still do not know. However, it is clear that vegetarians seldom get osteoporosis. Whether or not this is due to their greater intake of dietary boron remains to be seen. There could be other factors involved.

Chapter 19

BORON AND NEUTRONS

The major change in boron concentration in Drosophila occurred during the developmental stages. Eggs contained 82.6 ppm of boron on a dry weight basis. A dry weight basis means that the sample was dried out prior to analysis. Wet weight basis means that it was not dried out. Newly emerged adults contained 35.5 ppm boron. The boron content increased slowly and only slightly with aging, reaching a value of 56.9 ppm at 8 weeks of adult age. The fact that eggs contained the greatest amount of boron was interesting and suggested a possible reason for why boron is an essential nutrient. Perhaps it has a protective function.

An apple is really an egg produced by the apple tree with seeds to bring new life. We looked at apples and chicken eggs and found 11.7 ppm boron on a dry weight basis in the meat of an apple and 21.1 in the peel. The chicken egg had 18.0 ppm in the egg white, 7.3 in the yolk and 31.6 ppm boron in the eggshell. It is of some interest that the highest concentration of boron in both apples and chicken eggs is in the outermost section exposed to the environment. This suggests that the biological role of boron may be related to some protective function. This function could be related to both protection of the embryo and increased resistance to the aging process. It could be that boron functions as a protector against background radiation.

Environmental radiation

Artificially produced radiation at high levels in the form of x-rays, gamma rays, and neutrons is known to produce cancer and shorten the life span of most organisms. The effect of low levels of the numerous kinds of natural environmental radiation on life span, however, has received little attention except for the report of Giess and Planel (Influence de la Radioprotection Effectuee a Differents Stades sur la Longevite de Drosophila melanogaster, C. R. Acad. Sc. Paris, vol. 276, pp. 1029-1032, 1973) on the effect of lead shielding on the life span of Drosophila. Their surprising and unexpected result was that 10 cm (4 inches) of lead shielding actually <u>decreased</u> the life span of both male and female flies. This result is seldom referred to.

Lead is known to absorb the ionizing component of cosmic and terrestrial radiation. Environmental neutrons, however, which are produced by the interaction of cosmic radiation from space with the atmosphere, are affected very little by lead shielding. Therefore, the Giess and Planel experiment did not eliminate the possible effect of environmental neutrons on life span.

The question is: are neutrons involved in the aging process, and if so, how? The natural concentration of environmental neutrons is low compared to the total background radiation from all other cosmic and earth sources, with the flux (the constant exposure amount) being about 0.007 neutrons/cm^2 sec at sea level. This means that about every minute a human is exposed to about 5 neutrons per square inch.

The neutron flux increases dramatically with increasing altitude above sea level. Flying in an airplane will result in exposure to many more neutrons since they go right through the aluminum skin of the plane. Some environmental neutrons, found mostly at high altitudes, possess energies in excess of millions of electron volts, which means they will drill a hole through almost anything. Most of the neutrons found at sea level, however, are of low energy content and are referred to as thermal neutrons (0.025 electron volts or less at 20°C).

It is notable, however, that for neutrons, the lower the energy level the more biological damage is produced as measured by mutation induction or cancer production. This is probably because these neutrons move more slowly through biological tissues and, therefore, have more time to produce serious damage. These kinds of observations for the very low energy thermal neutrons suggest that they might possess considerable ability to induce biological damage. This damage might be reflected in the development of insect, animal, plant and human aging.

High-energy neutrons are known to reduce the life span of mice, even when given at relatively low doses. In contrast to the effects of other forms of radiation exposure, none of the damage as a result of neutron exposure appears to be repairable.

Protection against environmental neutron radiation
There are two known ways to protect against neutron exposure. One is by shielding with a high concentration of hydrogen atoms. Another, more effective way, is to use boron atoms as a shield.

Details

Paraffin wax is a commonly used moderator for slowing of neutrons due to paraffin's high hydrogen content and high capture probability. Collision of neutrons with hydrogen produces a large loss in neutron energy resulting in the production of low energy neutrons that can then be readily absorbed by the hydrogen atoms. The neutron absorption cross-section for hydrogen is 0.328 barns. The term "barns" is a measure of the ability of molecules to absorb neutrons.

The use of boron represents another effective means for capturing neutrons. The thermal neutron absorption cross-section for boron is an incredible 750 barns. This is a result of the 19.61% occurrence of [10]Boron (with an absorption cross section of 3990 barns) in naturally found boron. There are two kinds of boron in the diet just as there are three forms of hydrogen in ordinary water. These are called isotopes of the element. The point is that 3990 barns is better than 750 barns and a lot better than the 0.3 barns in paraffin wax.

Our theory was that boron was an essential nutrient because it gave protection against naturally produced environmental neutrons. We found mixed results for this theory. We found that shielding from environmental neutrons prolonged life span under some conditions but <u>reduced</u> life span when the shielding lasted through both the developmental and adult stages. This seemed to be a contradiction and required a lot of further experimentation.

When Drosophila adults were shielded from environmental neutrons with paraffin wax, there was an increase in life span. When shielding began at 1 day of age, the life span increased by 5.4%, but the increase was not statistically significant. Placing older flies of 14 and 29 days of age in the paraffin shield, however, gave significant increases in life span of 9.7 and 6.2%, respectively. Shielding from environmental neutrons with paraffin wax, therefore, improved the survival of Drosophila only for adult flies of 14 days of age or older.

Boron shielding, using bags of boric acid, confirmed the results found for paraffin shielding of environmental neutrons. Shielding begun at 1 day of adult age improved life span by 5.0% but, as for paraffin, the increase was not significant. However, shielding started at 27 days of adult age gave a similar increase in life span of 4.6%, which was significant.

One might ask the question how could a 5.0% increase in life span not be significant, but that a 4.6% increase could be significant? This is due to variation in the data. If one fly dies at an early age, it can throw the whole experiment off. The only thing that can make that experiment significant is if a number of other flies live to be older than normal. We, and many other laboratories, use statistical tests to confirm the validity of data. They are, in a sense, scientific lie detector tests.

The overall effect of neutron shielding in several experiments with both paraffin wax and boric acid was an improvement in life span for older flies only. This suggested a possible negative factor associated with neutron shielding which affected some process unique to younger flies. Therefore, we examined the influence of boron shielding on flies reared and maintained within the shield. Flies were allowed to produce eggs for 3 generations inside the boron shield before the newly emerged adults were placed in glass tubes for survival studies. Surprisingly, the median life span was reduced by 11.9% with the entire survival curve being shifted to reduced ages. This occurred when more than 99% of the environmental neutrons were removed by shielding with bags of boric acid.

Conclusion

Our results, therefore, confirmed and expanded upon those of Giess and Planel where they found a maximum reduction of the life span of Drosophila when the flies were shielded with lead for both the adult and developmental stages. Lead is effective for removing ionizing radiation but would have only limited effectiveness for absorbing neutrons. The thermal neutron absorption cross-section for lead is only 0.17 barns. It seems unlikely that the reduction in life span as a result of lead shielding was due to neutrons. Our results indicate that the removal of normal neutron exposure also has a negative influence on life span. This study and that of Giess and Planel force us to conclude that background radiation in general must have some positive influence on longevity when exposure occurs during both the developmental and early adult stages of Drosophila.

Low-level radiation exposure

However, removal of background neutron exposure for older flies of 14-29 days of adult age has a small beneficial effect on life span. The well-known damaging effects of neutrons, such as chromosome breakage, must, consequently, be a factor during the later stages of adult life. We must draw the unlikely conclusion that some, as yet unknown, process is activated or improved during the earlier stages of life by environmental neutrons, resulting in greater life span.

This is not an original proposal. Henry, H.F (Guest Editorial: Is all Nuclear Radiation harmful? Health Physics, vol. 43, pp. 767- 769, 1982), Luckey (Physiological Benefits from Low Levels of Ionizing Radiation, Health Physics, vol. 43, pp. 771-789, 1982) and Congdon (A Review of Certain Low-level Ionizing Radiation Studies in Mice and

Guinea Pigs, Health Physics, vol. 52, pp. 593-597, 1987) have all proposed that low-level radiation is of benefit to organisms with Luckey stating that "radiation may be essential for life." Luckey has also reviewed numerous reports of increases in longevity in a great variety of species as a result of exposure to minute doses of radiation.

To my knowledge, however, only we and Giess and Planel have reported decreases in longevity as a result of shielding from background radiation. This and the reports reviewed by Luckey suggest that perhaps an optimum radiation exposure level exists for each organism that could lead to maximum life span. A possible molecular basis for such an effect has already been demonstrated in Drosophila by Kennison and Ripoll (Spontaneous Mitotic Recombination and Evidence for an X-ray-inducible System for the Repair of DNA Damage in Drosophila melanogaster, Genetics, vol. 98, pp. 91-103, 1981) who found that x-rays appeared to induce a repair system for DNA damage.

Other forms of radiation may have similar effects. Human lymphocytes (white blood cells) have also been shown to become resistant to x-ray damage after being cultured in the presence of thymidine containing tritium. This acted as a chronic source of low-level radiation. Tritium is a mildly radioactive form of hydrogen which occurs at low concentrations in most water samples, and thymidine is a precursor for the synthesis of DNA.

In conclusion, it is possible, but seems unlikely, that all of these studies, including our results with fruit flies, can be ascribed to experimental errors. However, it is important that additional studies be done at various low-level radiation intensities in other organisms before definitive statements can be made regarding the influence of background radiation exposure on longevity.

It has been proposed that the improvement in longevity

as a result of low-level radiation exposure may represent a more general phenomenon elicited by nonspecific stimuli. The term "hormesis" has been proposed to describe the favorable response, including increases in life span of living organisms, to a wide range of toxicants when given at low concentrations. These include radiation, pesticides, carcinogens, organic solvents and mild physiological and sensory stresses. It is possible that background radiation exposure represents an example of "longevity hormesis."

Neutron exposure and where you live
Whether or not the environmental neutron flux affects the life span of organisms other than fruit flies remains to be established. The neutron density and energy levels change considerably with altitude with approximately a 10-fold increase in going from sea level to 5,000 meters, for example. While we did not have the capability for measuring neutron flux, our laboratory was located at an elevation of 150 meters and would, therefore, have had a relatively low neutron flux. A laboratory located at a higher elevation might, consequently, find differences in life span much greater than those found at our location for the same amount of shielding.

Neutron flux also changes with geomagnetic latitude. People living near the equator at sea level have very little neutron exposure, whereas those living closer to the North or South Pole have the maximum. People living in Panama have a lot less neutron exposure compared to persons living in Iceland. It is of interest that Iceland has one of the world's greatest life expectancies.

At the elevation and latitude (43°N) of our laboratory, the cosmic-ray neutron dose equivalent rate according to Nakamura, T., Uwamino, Y., Ohkubo, T., and Hara, A. (Altitude Variation of Cosmic-ray Neutrons, Health Physics, vol. 53, pp. 509-517, 1987) should be 57 µSv/year (µSv is a measure of radiation exposure). This is a small percentage of the total estimated "natural" radiation exposure due to other kinds of radiation of 1003 µSv/year from all sources (cosmic, terrestrial and internal radiation due to incorporated radionuclides). Quantitatively, with a flux of 0.007 neutrons/cm² sec at sea level, each square centimeter of surface area of an unshielded organism would be exposed to 604 neutrons/day. This would be the approximate flux rate for neutrons at the altitude and latitude of our laboratory location. Our results suggest that this relatively low exposure may have both negative and beneficial influences on biological organisms, depending on their age.

The space programs and life span

Another factor affecting neutron flux in the future is the predicted reduction by a factor of 10 in the lifetime of protons in the Van Allen proton belt above the earth by the year 2010 due to all the space debris introduced by the American and Russian space programs. Since environmental neutrons result from the interaction of cosmic rays with air and protons, this means that the number of neutrons reaching the surface of the earth will be greatly reduced. Such a change could have profound consequences for many biological processes, including aging. This could be a serious problem if neutrons are necessary for life. On the other hand, if low levels of neutrons are harmful, then we may see a benefit in terms of human life span. The question of how this might affect the rate of aging in humans and other organisms, however, remains unanswered. In a sense this is a large scale experiment which all are subject to, like it or not.

Boron and radiation, still a puzzle

The boron content of biological systems may also be a factor in the response to environmental neutrons. The very high neutron capture cross section for boron-10 has been exploited for cancer therapy based on the production of an alpha particle and gamma-ray energy release as a result of neutron capture by natural boron-10 in tumor tissues. Boron is an essential nutrient but with no known biological function. Since the abundance of boron-10 in boron compounds is 19.61%, natural boron also has a very high neutron capture cross section.

We have previously reported that the boron content of fruit flies is very high during the egg stage and then declines rapidly during the subsequent developmental stages. The boron content of adult flies, however, increases with age. These changes and the cellular location of the boron may affect how the organism responds to environmental neutrons during the different stages of life. The improved survival of older flies shielded from environmental neutrons may be due to their greater cellular content of boron, which might normally capture neutrons, and thereby produce cellular damage. This damage would be caused by the release of alpha particles and gamma rays as a result of the neutron capture. We are, however, unable to explain why the developmental and early adult stages of life would benefit from background neutron exposure.

Overall, it is clear that low levels of radiation have a beneficial influence on life span. Most people would find this to be an outrageous statement, but the evidence is considerable. Dietary boron appears to be somehow involved in this process. Our original theory that high levels of boron in egg shells and apple peels serve as a protection against environmental neutrons has not been proven. It is possible that neutrons are essential for life and that too much boron in the form of shielding or diet may reduce them to an undesirable level. This could be the basis for boron toxicity in both plants and animals. On the other hand, excess neutron exposure as the result of boron deficiency, could lead to a greater rate of aging. If boron does act as a shield against environmental neutrons, then human skin should contain the greatest amount of boron. I wish that we had done such a measurement. These issues have not been resolved and deserve at least some scientific attention.

Chapter 20

LIGHT

As an introduction, it should be said that there are three forms of light, ultraviolet, visible and infrared. Humans see only visible light. Ultraviolet light has the most energy and, for this reason, it causes skin cancer and other skin problems. Bees can see this kind of light, but people cannot. Infrared light is the weakest form of light. We cannot see it except with special equipment. Anything that gives off heat also gives off infrared light, so with this equipment one can see people, or anything else which gives off heat, in the dark. Sunlight is composed of all three forms of light. Man-made light, used to light homes and businesses, is almost entirely visible light. Consequently, most of the light humans are exposed to is visible light.

"Lux" is a measure of light exposure. On a cloudy day or in the shade the normal exposure is about 15,000 lux, but in full sunlight it can be more than 100,000 lux. Light is measured in photons, which are the elemental particles of light, with "lux" being a way of expressing the number of photons. If a person were outside for 60 minutes at 15,000 lux, the total exposure would be about seven times less than it would be at 100,000 lux for the same amount of time. There are all sorts of terms for light exposure with "lux" being the easiest way of expressing it.

Fruit flies

Northrop first reported the influence of light on the duration of life of fruit flies in 1925 (The Influence of the Intensity of Light on the Rate of Growth and Duration of Life of Drosophila, J. Gen. Physiol., vol. 9, pp. 81-86). While the nutritional conditions used in his experiments would not be considered adequate today, he found very large changes in life span at 20, 25 and 30°C (degrees Celsius) for light intensities ranging from 0 to 10,000 lux. Surprisingly little work has appeared since then. Incidentally, Northrop went on to win the Nobel Prize for some other work that he did.

More recently, Allemand reported that permanent darkness had a much greater effect on the life span of Drosophila than did changes in light cycles (the amount of light given at different times of the day). Female flies lived 43% longer at 25°C in permanent darkness than in the presence of visible light at 1500 lux which is not very much light. Erk and Samis also found very large increases in life span for male and female Drosophila when kept in varying periods of daily darkness as compared to constant visible light.

Exposure to ultraviolet light, which is more energetic than visible light, has also been shown to result in a dramatic reduction in the life span of Drosophila.

Changes in life span have also been observed as a result of changes in light cycles, called circadian cycles, in house flies, blowflies and rotifers. In all cases, however, those organisms receiving the greatest amount of light had the shortest life span regardless of how these light cycles were changed. Thus, it seems that the amount of light, rather than the cycles, changed the life span.

In an effort to separate circadian effects from those due solely to light exposure, we examined the influence of visible light on the life span of flies exposed to a constant 12 hours on and 12 hours off light cycle. We also directed our studies to higher temperatures where photochemical processes would be enhanced. A photodynamic process would predict even greater changes at higher temperatures than those previously reported for fruit flies at 20 and 25°C.

Visible light is the most common form of electromagnetic radiation to which most organisms are exposed. Most of us do not think of visible light as being a form of radiation, but it is. It is a low energy form of radiation, and we are exposed to a lot of it compared to other forms of radiation such as x-rays. Sunlight on a typical cloudless day is usually in excess of 100,000 lux. Even in full shade the outdoor value is 15,000 lux.

High intensities of light can produce heat in objects exposed to such radiation. Since it is well known that the environmental temperature affects the life span of fruit flies, any study of the effect of light on life span must exclude possible temperature changes. Our studies were, therefore, limited to relatively low light intensities where the temperature remained constant. We would have preferred to have looked at the effect of high light intensities, but could not because of possible increases in temperature due to the intense light.

Studying the effect of visible light at higher temperatures allowed us to determine changes in life span larger than those previously reported. At 30°C we found the median life span to be 34 days under "dark" (4 lux) conditions. Exposing the flies to different intensities of light for 12 hours/day decreased life span in a dose dependent manner. At 7,300 lux, for example, the life span was reduced to 14.5 days with the entire survival curve shifted to lower values. Simply changing the amount of light exposure from 7,300 to 4 lux, therefore, increased the median life span by an incredible 141%. It should be mentioned that 7,300 lux is not very much light when you consider that the amount of light from the sun during the day in the summer is about 100,000 lux.

At 35°C even more dramatic changes were observed. The median life span in the dark (0.3 lux) was 63.6 hours, and 5.0 hours at 6900 lux. This amounts to an amazing 1172% increase in life span just by putting the flies in the dark. This was the greatest increase in life span observed in any of our experiments. In another experiment at this temperature, but with less of a difference in the light exposure, the increase in life span was 389%. Even very low light intensities resulted in considerable reductions in life span. At 65.5 lux the median life span was reduced to 44 hours. Again it should be stressed that the flies were maintained on a 12:12 hour light/dark cycle. The effect was, therefore, photochemical in nature rather than due to diurnal (time of day) influences.

At 37°C the median life span was greatly reduced, due to the elevated temperature, to 8.85 hours under dark (0.3 lux) conditions. Adding light exposure at 6500 lux, however, reduced the median life span even more to 1.60 hours, which means that just placing the flies in the dark increased their life span by 453%. Even very dim light of 60.0 lux reduced the median life span to 8.0 hours at this temperature.

Other evidence

While visible light is generally regarded to be harmless, there are many reports in other model systems indicating biological toxicity. It has been shown, for example, that 3 hours of fluorescent light at a low dose of 1,000 lux can induce more strand breaks in the DNA of mammalian cells than a very large dose of x-rays (M. O. Bradley, L.C. Erickson and K.W. Kohn, Non-enzymatic DNA Strand Breaks Induced in Mammalian Cells by Fluorescent Light, Biochim. Biophys. Acta, vol. 520, pp. 11-20, 1978). I have referenced this paper because it is an almost unbelievable report, since it is known that a single dose of 300 rads of x-rays can greatly increase the rate of aging in mice and rats, reducing their life span by almost 50%. This is a most disturbing result and has to make us all wonder about fluorescent light bulbs. It is not, however, the source of light which is probably the problem but light itself. The reason for saying this is that other sources of visible light also produce serious problems.

For example, ordinary light bulbs, usually called incandescent lamps, (which lack ultraviolet and most of the blue light) are found in most homes. This light source can produce what are referred to as "pyrimidine dimers" in the genetic DNA molecules. These dimers are paired-up molecules and represent considerable DNA damage. The most frequently found, as a result of exposure to light, is the thymine dimer.

Other forms of visible light also produce toxic products in bacteria and in animal cell culture media. Repair in viruses and bacteria is inactivated by visible light, and bacteria experience mutagenesis when exposed to visible light. This could be the real reason we have more colds and flu in the winter months, since sunlight kills bacteria and viruses.

Low intensities of visible light also inhibit protein synthesis in yeast, and damage has been observed in liver cells exposed to visible light. Production of superoxide radicals has been induced by visible light in whole cells of algae, and light-induced free radical oxidation of membrane lipids has been reported in the photoreceptors of frog retina. DNA-to-protein crosslinks have been induced in mammalian cells with visible light. This is an undesirable situation since both the DNA and the protein become dysfunctional. It has even been proposed that the extensive cross-linking, or abnormal connecting together, of molecules of this nature is a basic cause of aging. This is called the cross-linking theory of aging.

The ability of light to penetrate the body is well known to anyone who has placed a flashlight under their hand. Penetration of visible light into the brains of mammals has been measured, with smaller mammals experiencing the most internal light exposure. Visible light can apparently even pass through the skull. Perhaps, this is another reason why smaller mammals generally age faster than larger mammals. The wide variety of light-absorbing molecules in any biological system makes it unlikely that any particular molecule is responsible for visible light-induced damage. Photodynamic action is defined as "the oxidation of biologically important molecules in the presence of oxygen, a dye, and visible light." In fruit flies and other organisms certain natural molecules may function as dyes. In the presence of light and oxygen, oxidation reactions probably occur. Increasing either the oxygen concentration or the temperature would result in more extensive photo-oxidation. These two factors of increased temperature and oxygen are, by themselves, known to increase the rate of aging of fruit flies independent of any changes in light exposure.

A major factor in the aging of fruit flies and other organisms may be photochemical processes similar in nature to photodynamic action that involves the presence of oxygen, light and natural dyes. Visible light is by far the most common and intense form of radiation exposure experienced by most organisms. The results with fruit flies suggest that additional studies with other organisms may provide a better understanding of how light affects aging in general.

Aging is an, as yet, undefined phenomenon that affects living organisms at greatly different rates. Aging is not a disease but rather a fundamental biological process involving biological, chemical and physical factors. Exposure to light is one of the physical processes.

The effect of visible light on biological systems has been studied in two basic areas. Chronobiology ("chrono" means time) has been concerned with the influence of cycles and rhythms on a wide range of biological and biochemical changes. Many of these changes have been altered by changing the timing of the inducer of the cycle. The "time giver" has often been the presence or absence of visible light. Moderate to very large changes in the activities of enzymes, catecholamines, levels of hormones and function of the immune system have been reported for different times in the light-dark cycle.

Some of this knowledge has been exploited for improved chemotherapy of tumor growth based on the differences in the changes in the circadian rhythm of normal and cancerous cell division. In short, you can kill many more cancer cells at certain times of the day with radiation or drugs with less harm to normal cells. Even the concentration of copper and iron in human serum undergoes variation during the course of a 24 hr period in the presence of a light-dark cycle. It is clear, therefore, that the changes in light and dark can produce considerable changes in a wide range of biological parameters.

The second area of light research has involved studies of how different wavelengths and intensities of light affect biological systems. It has been clearly established that ultraviolet light (this is light with the greatest energy) produces a wide range of changes, including DNA damage and tumor induction. Humans, however, have a clearly proven need for an exposure of at least 10 minutes per day of ultraviolet light in order to produce vitamin D. It seems, nevertheless, that dietary intake by eating fish or dairy products would be a better way to do this. Overall, it is clear that visible light, while popularly considered benign, has been shown to produce a wide range of biochemical changes of a negative nature.

Photosensitizers

The mechanism of photoaging at present is unknown, but there is some research that suggests a possible explanation. Photosensitizers are small molecules that increase the effect of visible light on organisms. For example, people taking certain antibiotics are advised to avoid sunlight because these antibiotics are photosensitizers. In an effort to elucidate the process of photoaging, we examined the effect of a photosensitizer, ethylene blue, which is a blue dye, on the life span of fruit flies kept at 25 degrees Celsius.

Placing young adult fruit flies on medium containing a small amount of ethylene blue gave a 49% reduction in the median life span with the entire survival curve being shifted to shorter survival times. This large reduction in longevity occurred with a daytime light exposure of only 1000 lux, which is the amount of light in a typical home with the shades down. I believe that if we had exposed them to full sunlight that they would have died in less than a single day from the intense light exposure and the presence of the photosensitizer.

The result with ethylene blue suggests that photodynamic action which requires the presence of oxygen, light, and a photosensitizer, may be an important mechanism in the basic aging process. It is generally accepted that the energy of visible light is transferred to the photosensitizer. This photosensitizer in turn transfers it to molecular oxygen. One could think of this as passing the ball around in a basketball game. The result is the production of more reactive oxygen species involved in oxidative processes called photodynamic action. This is really an energy transfer process starting with the light exposure. The chemists have worked this all out in great detail for chemical reactions.

It is known, thanks to the chemists, that photosensitized oxygenation of ethylene blue results in the production of singlet oxygen. We have also previously shown that exposure to singlet oxygen increases the rate of aging of Drosophila (Chapter 9). It is reasonable to propose, therefore, that the greatly increased rate of aging produced by ethylene blue is due to the production of the very reactive molecule, singlet oxygen.

It should be said that not all photosensitizers reduce life span. We found, for example, that the vitamin riboflavin, which is a photosensitizer, can prolong the life span of fruit flies under certain conditions of light exposure.

Mammals and birds

The current evidence for photoaging in mammals is limited, but several observations suggest that environmental light intensity may affect the life span of mammals. The 1975 report from Germany of Kunstyr and Leuenberger (Gerontological Data of C57BL/6J mice. Sex Differences in Survival Curves, J. Gerontol., vol. 30, pp. 157-162) dem-onstrated greater longevity for C57BL/6J mice when kept under conditions of low light. Their mice were exposed for a lifetime to 55 lux for a few hours per day and remained at 0.3 lux for the rest of the time. They were, therefore, in the dark most of the time. The median survival time of 917 days for 296 male mice was by far the greatest life span reported for C57BL/6J mice up to that time, almost twice that of some other laboratories. They found a simi-lar result for female mice. Their oldest mouse lived to be 1,264 days which is most unusual, even today.

Unfortunately, these scientists did not compare different light exposures, but this was not the point of their experi-ment. They were looking at the differences in life span be-tween male and female mice. The male mice lived slightly longer. At the time, the authors ascribed their unusual re-sult to a "genetic drift" which they proposed to have oc-curred between 1966 and 1970, since their mice were born on October 30, 1967. There is, however, little evidence for such a proposal. The low light exposure seems a more likely reason.

A report by L.E. Scheving and J.E. Pauly (Several Prob-lems Associated with the Conduct of Chronobiological Research, Nova Acta Leopoldina, vol. 46, pp. 237-258, 1977) indicates that rats experience a wide range of prob-lems when exposed to continuous light of 650 lux, which is not very much light, beginning at 3 weeks of age. Rats generally come out only at night, so even 650 lux is more light than they are usually exposed to. At 32 months of age the rats were poorly groomed with dirty, yellowish and matted hair. Upon autopsy "there was no thymus, al-most no subcutaneous or mesenteric fat and no brown fat to speak of." These rats also had opaque eyes and sores on the tails and sides of the body. Rats kept in the dark, however, had a "considerable amount of subcutaneous fat" and "had a good size thymus for an old rat." They had no sores. The photographs of these animals show a dramatic improvement in overall appearance compared to that of the rats kept in continuous light. Professor Lawrence Scheving has communicated to me that while his experiment was not a survival study, the rats kept in the dark lived longer than those kept in constant light. He also told me that those on a 12:12 hr light/dark cycle had a survival intermediate between those in constant darkness or constant light.

Another indication of the possible effect of visible light on mammals involves the observation of unusually long life span for bats. Austad and Fischer (Mammalian Aging, Metabolism, and Ecology: Evidence from the Bats and Marsupials, J. Gerontol., vol. 46, pp. B47-53, 1991) conclude that "bats have maximum life spans a minimum of 3 times those of nonflying eutherians--a trend resulting from neither low basal metabolic rate, the ability to enter torpor, nor large relative brain size."

Bats, therefore, have none of the characteristics associated with increased life span in mammals, yet they have achieved exceptionally long life spans for mammals of their size. Since bats are known to have good eyesight, their nocturnal existence seems to be of choice, rather than necessity. The phrase "blind as a bat" is simply not true. The observed life spans of bats ranging from 7 to 30 years of age are exceptional for mammals of such small size, especially when compared to a mouse or rat of the same size which live only 2 to 3 years of age. This is truly an exceptional difference. At present there is no adequate explanation for this observation except for the possibility that the differences are due to variations in light exposure.

It is possible that birds live much longer than mammals

of the same size, even though they have a higher body temperature, because of their almost complete coverage with feathers. Perhaps feathers are better light protectors than fur. In addition, birds of very great longevity, such as the owl, are nocturnal.

Humans

Humans are very large compared to most organisms, and it would take a lot more light to penetrate deeply into body tissues than it would take, for example, for a fruit fly. Skin cancer is by far the most common form of cancer in humans and is widely believed to be due to exposure to ultraviolet light. It is much more prevalent in areas of the world, such as Arizona and Australia, where people are exposed to higher than average amounts of sunlight. It is possible that both visible and ultraviolet lights are responsible for this condition, and not just ultraviolet light alone. There is some recent evidence for this proposal. In general, people who spend a lot of time in the sunlight do not look very good as they grow older. It seems that excess exposure to light is a problem, but it is not clear whether or not it is as serious a problem for humans as it is for fruit flies.

Until recently the oldest living man was reported to be a Welsh coal miner who continued to do hard work even into his 90's. This man was probably exposed to less sunlight than most of us. I remember that women used to wear hats and white gloves whenever they went out into the sun. Fishermen who wear sunglasses seldom develop cataracts of the eye, whereas those who do not use them have a very high incidence of this problem. In certain areas of the world where sunlight exposure is great, people wear heavy protective clothing even in the heat of summer. They must do this sort of thing for a reason.

The human body temperature is high, 37 degrees Celsius, which is one of the temperatures where we found a rapid rate of photoaging in Drosophila. It is highly unlikely, however, that a more than 400% increase in life span, as we observed in fruit flies at this temperature, could be achieved in humans by reducing exposure to visible light. That would amount to an increase of about 300 years in the average human life span. This is very hard to believe and probably is not a possibility. The potential for a large increase in human life span, however, exists.

Plants
Probably the best approach to this problem is to study plants. Plants are exposed to very high levels of visible light on a constant basis, yet they seem to suffer no negative consequences unless man gets in the way. Farmers use herbicides to control weeds, and one of these used to be 3-AT. This herbicide inactivates the enzyme catalase. Catalase removes the oxidizing molecule hydrogen peroxide by converting it to oxygen and water. Without catalase the plants oxidize, wither and die as a result of light exposure. It is reasonable to propose that catalase also protects humans against light exposure. This is one of the important enzymes that declines with aging. The chemical 3-AT is not used anymore in agriculture because it causes cancer of the thyroid in humans.

There are likely hundreds of other molecules which protect plants against the effects of visible light and, perhaps, some more powerful than catalase. Catalase, however, is the one enzyme that we know protects plants against light exposure. Reducing the level of catalase with 3-AT changes the life span of Drosophila. A concentration of 1% 3-AT in the food decreases the life span of fruit flies by 65.2% under normal light exposure, and, even 0.001%

3-AT reduces life span by 6.0%. The real function of the enzyme catalase in organisms including plants, animals and humans may be to protect us from the damaging influences of visible light.

Plants, however, must contain numerous other, more effective, molecules for light protection. If it is considered that a Bristle Cone pine tree can live to be 5,000 years of age, there must be molecules of which we know nothing which confer great protection against the basic aging process and possibly those induced by light. Unfortunately, little or no research has been done on this subject.

The evidence for protection against cancer and cardiovascular disease through the consumption of a high plant-containing diet is considerable. It is also known that vegetarians live about 9 years longer than most people, but this is not very much when you consider the great potential for the improvement in human life span by, as yet unknown, photoprotective dietary agents. In conclusion, we are really in the dark ages concerning this whole subject.

Reduced sunlight exposure is a real problem for many people who suffer from depression, especially during the winter months. Since we really do not know if reduced light exposure prolongs the life span of humans, it would be ill advised for these people to reduce their exposure to light.

Chapter 21

EXERCISE AND LACTIC ACID

Oxidation reactions mediated through free radical mechanisms have been implicated in numerous disease states and the aging process itself. Regulation of these destructive oxidation pathways by the use of various antioxidants offers the potential for the control and prevention of degenerative disease states and biological senescence. Examples of well-known natural antioxidants include vitamin C and vitamin E. Lactic acid, which has generally been considered to be only a metabolic waste product, also serves as an important antioxidant in man and other organisms.

Antioxidant enzymes have been shown to play a vital function in the maintenance of life forms of all kinds. Catalase, glutathione peroxidase and superoxide dismutase have all been shown to be essential for antioxidant protection. Other proteins such as ceruloplasmin and metallothionein have also been proposed as significant cellular antioxidants. Smaller molecules such as ascorbic acid, vitamin E, vitamin A, uric acid and glutathione have also been shown to have powerful antioxidant functions.

The primary free radical or oxidant molecules involved

in biological damage appear to be lipid peroxides, peroxy radicals, hydrogen peroxide, superoxide radicals, singlet oxygen, and hydroxyl radicals. All of these species have relatively long half-lives (that is how long that they exist before disappearing), except for the hydroxyl radical. Singlet oxygen, for example, can undergo more than 600 molecular collisions in a typical biological system before a reaction occurs with another molecule. The highly reactive hydroxyl radical, on the other hand, reacts at essentially a diffusion-controlled rate which means that it combines with the first thing it encounters. An analogy could be that the hydroxyl radical is highly promiscuous, whereas singlet oxygen is much more selective.

How can these really harmful hydroxyl radicals be trapped? In order for a hydroxyl trap to be effective it must: (1) be inducible under oxidative stress; that is when free radicals are being produced; (2) be at relatively high concentrations and (3) be in close proximity to the site of formation of the hydroxyl radical. Ideally, the product of the trapping reaction should be a non-toxic compound of little or no reactivity with other molecules. Lactic acid is such a molecule. Milk contains the sugar lactose that can be converted to lactic acid if fermented by the bacteria that produce yogurt, sour milk, and buttermilk. These are good sources of dietary lactic acid.

Lactic acid as an antioxidant
Several lines of evidence suggest that lactic acid may serve as an effective antioxidant. Lactic acid is listed as a GRAS (generally regarded as safe) food additive and has been used as such for hundreds of years. Historically, it has been used as a food preservative. It has been shown, for example, that lactic acid addition protects beta-caro-

tene from oxidation in cottonseed meal and other grain mixtures.

Under certain conditions the salt of lactic acid, sodium lactate, can protect cells in culture from potentially lethal x-ray radiation damage. This is consistent with the proposal that lactic acid is acting as a hydroxyl radical-trapping agent since it is well known that the primary damaging agent produced by x-ray radiation is the hydroxyl radical.

It is now generally accepted that ischaemia (lack of oxygen), which occurs during a heart attack or stroke, leads to elevated free radical production. It has also been recently reported that lactate protects synaptosomes (nerves) from cell death and calcium overload under ischemic conditions, which occur during a stroke.

The mechanism for lactic acid protection appears to result from the reaction of hydroxyl radicals with lactic acid as first reported in 1898, by Ruff who showed that the reaction products were carbon dioxide and the next lower sugar (O. Ruff, Ueber die Verwandlung der d-gluconsaure in d-arabinose, Ber. Ges. Physik. Chemie, vol. 31, pp. 1573-1577). He found this to be true for all aldonic acids (acids of different sugars) including lactic and gluconic acid. It is truly incredible that these early research scientists were able to do what they did with so little information available at the time to help them.

Exercise is known to result in greatly elevated concentrations of lactic acid in the blood and is perhaps the most effective means for increasing lactic acid levels. Retzlaff and Fontaine reported some years ago that exercising rats for as little as 20 minutes per day increased their life span by 11% (E. Retzlaff and J. Fontaine, Functional and Structural Changes in Motor Neurons with Age, In: <u>Behavior, Aging, and the Nervous System </u>(Edited by A.T. Welford and J.E. Birren) Charles C. Thomas, Springfield, Illinois, pp. 340-352, 1965). Spinal motor cells in the exercised group at older ages were similar in appearance to those seen in young animals. This was in spite of the fact that the exercised rats gained considerably more weight (This might have been due to an increase in muscle mass.).

More recently, epidemiological studies have indicated a similar result for humans with a lower mortality from all causes in men who exercise regularly. Numerous factors change as a result of exercise such as improved glucose tolerance and insulin sensitivity, higher levels of the beneficial high-density lipoprotein cholesterol, and improved oxygen uptake. Associated with this increased oxygen uptake is an exponential increase in blood lactate, even under aerobic conditions (when you are getting lots of oxygen). It seems that lactic acid is required as an antioxidant to prevent the oxidative damage now known to be associated with physical activity.

Exercise and immune response
In addition to its proposed antioxidant function, lactate may have a regulatory role in other biological processes. Moderate exercise is known to enhance immune function in several different ways, including elevated antibody production in mice. In humans, moderate exercise increases the lymphocyte stimulation index, natural killer activity and delayed hypersensitivity. These are all good things for the immune system. Physical activity also increases plasma growth hormone and prolactin concentrations.

Amazingly, infusions of the salt of lactic acid, sodium L-lactate, accomplish the same effect as exercise. Lactate has also been shown to regulate interleukin-2 (this is an important immune system molecule) production in T-lymphocytes at the concentrations of lactate achieved in normal exercise. The induction of interleukin-2 in these studies was substantial, especially in the CD4, T-lymphocyte cells. These are the cells that people with AIDS lose with disastrous consequences. The mechanism of this effect is unclear but is associated with reduced concentrations of intracellular glutathione. What this all means is that lactic acid, derived from exercise, is good for the immune system and will probably protect against infection and cancer.

Regular intense heavy-duty exercise training programs, however, are actually immunosuppressive. They can make athletes sick, which may be due to excessive lactic acid production. This is probably why marathon runners often develop colds after the run.

The amount of exercise or lactate required for optimum immune function at present is not clearly established, especially for different age groups. It is noteworthy that the amount of lactic acid produced during exercise in even highly trained human athletes declines considerably with age. It is clear that much of the exercise-induced changes in the immune system are mediated through lactic acid production.

The possible involvement of lactic acid in immune response would seem to be of potential usefulness in efforts to restore the decline in immune function known to occur with aging. Specifically, interleukin-2 production and T-lymphocyte response have both been shown to decline greatly with aging. These are very important components of the immune system which protect against cancer and infection. This has been observed by many investigators and has been graphically documented. Antioxidant status as reflected by the levels of molecular and enzymatic antioxidants in individual organisms could represent an important factor in the level of immune response. I believe that the evidence shows that lactic acid is one of these antioxidants, if not the major one.

Both the positive and negative effects of exercise on immune function and longevity can be explained by changes in lactic acid concentrations. Since moderate exercise results in low levels of lactic acid production, low concentrations of lactic acid could function as an effective antioxidant and activator of immune function. On the other hand, very intense exercise results in large increases in lactic acid concentration and decreased immune response. These very high lactic acid concentrations change something, but I do not know what it is. Consistent with this proposal is the observed protective effect of low concentrations of lactate against radiation damage, whereas high concentrations, surprisingly, enhance cell death from radiation exposure. This is like the boron story. Too little boron reduces life span, and too much also reduces it. It takes just the right amount of boron to prolong life span. It seems important to also establish what the optimum lactic acid concentration needed for maximum immune function should be in humans and other organisms.

It has been established that the enzyme, lactic dehydrogenase, declines with age in rat heart tissue. This seems of special interest in view of the preference of heart muscle for lactic acid as a fuel source. This could be another reason why moderate exercise is so beneficial for the prevention of cardiovascular disease. Lactic dehydrogenase activity also declines with age in human aortic and pulmonary blood vessels, in mouse kidney, in rat brain and in mouse skeletal muscle. Other tissues show increases or no change with age.

It is well known that exercise results in elevated lactic acid concentrations in the muscles and in the blood. Retzlaff and Fontaine have reported that exercising rats, as noted above, for as little as 20 minutes per day increases their life span by 11%. In addition to increased life span, the exercised rats showed enhanced spinal motor neuron integrative capacity and increased motor axon large fiber conduction velocity. This means that their nerves were greatly improved. Why there is more lactic acid in nerve tissue than any other organ is unknown. It probably has a protective function against free radicals. If increased concentrations of lactic acid were responsible for the results of Retzlaff and Fontaine, then dietary supplementation with lactic acid might be expected to achieve the same life prolongation that they found with exercise.

Dietary lactic acid

We found that feeding lactic acid increases the life span of the fruit fly. The normal lactic acid content of Drosophila reaches a peak during the late pupal developmental stage and then declines by 45% during the first two weeks of the adult stage. Feeding lactic acid during the developmental and adult stages increased the median life span by 15%. Feeding during only the adult stage increased life span by 12%.

Rearing and maintaining flies on another aldonic acid, gluconic acid (the sugar acid of glucose found in honey wine), increased life span by 22%. Feeding gluconic acid during the adult stage increased life span by 12%. Gluconic acid is another officially approved food preservative that is used mostly in summer sausages. It is a larger molecule than lactic acid but chemically behaves almost the same. Overall, our results indicate that lactic and gluconic acids prolong the adult life span of Drosophila but only if given early in life and continuously.

It should be pointed out that the favorable effect of dietary lactic acid on longevity occurred only at lower concentrations. At higher concentrations we observed a reduction in overall survival. This is similar to the reduced immune response and greater x-ray damage reported at higher concentrations of lactic acid in humans and human cells. One might expect that excessive exercise would do the same thing, since it produces very high levels of lactic acid.

Aging and lower lactic acid levels

Our unpublished studies indicate that lactic acid concentrations decline with age in the mouse. We found that the lactic acid content of mouse blood decreases by more than 50% between 30 and 900 days of age. The decline was continuous up to the very old age of 900 days.

Since lactic acid concentrations are generally known to be the highest in nerve tissues, we also examined the lactic acid content of mouse brain. Again, there was a significant decrease of 21.5% with age between 30 and 900 days.

These results are in agreement with the decline in lactic acid concentrations seen with aging in Drosophila. It seems unlikely that the changes in mice are entirely due to reduced physical activity in old age, since we found no change in the lactic acid content of mouse blood on a diurnal basis (time of day basis). Mice are somewhat more physically active at night, so you might expect to find higher levels of lactic acid at night. We did not find this to be true. The mice were sampled every 4 hours for 24 hrs on a 12:12 light/dark cycle. These were middle-aged mice (460 days of age). It is possible, however, that diurnal changes occur in other age groups. Overall, it appears that some intrinsic process is causing a decline in lactic acid levels with age. It could be due to something as simple as a lack of enough exercise or possibly because of the reduced oxygen consumption that normally occurs with aging. Mice, like most people, tend to avoid exercise as they grow older. This is probably why there is a 20 to 40% decline in muscle mass in older people. This decline in muscle mass could be the reason for lower lactic acid concentrations in older organisms.

Lactic acid and differences in life span

It also appears that lactic acid concentrations reflect the metabolic rate in different species. The lactic acid content of human blood is on average 7.5 mg/100 ml, whereas that of mouse blood is 30.0 mg/100 ml. We estimate that whole Drosophila contain more than 200 mg/100 ml of lactic acid. For humans and mice, these numbers are roughly in proportion to the differences in the metabolic rate of these two species. The 7-fold higher basal metabolic rate for mice compared to man is reflected in a 4-fold higher lactic acid concentration in mouse blood. This is not the same but pretty close. A similar relationship holds for fruit flies where the basal metabolic rate is 13.3 times higher than that for a human. Do these organisms have a shorter life span than humans because of higher levels of lactic acid? It is more probable that lactic acid protects them against high levels of oxygen consumption.

Lactic acid seems to be involved in the basic aging process. Higher levels occur in organisms that age more rapidly. The question is whether or not this is a protective mechanism. It seems that it is because increasing the dietary intake of lactic acid prolongs life span, at least in fruit flies. Very high dietary lactic acid, however, reduces the life span of fruit flies. Physical exercise increases the production of lactic acid in rats, and this is associated with an increase in life span. However, excessive physical exercise is known to reduce immune response in humans.

There is no doubt that moderate exercise improves the health of humans, but what the appropriate level is remains to be established. There is little doubt that intense and repeated exercise is a problem in terms of reduced immune response in humans.

Many people cannot exercise due to physical and other limitations. Dietary lactic acid is a possible solution to this problem. Yogurt, buttermilk and other fermented milk products, such as sour milk, contain large amounts of lactic acid. The reason for this is that the sugar lactose that occurs in milk is converted to lactic acid by bacteria added to these products. There are claims of great longevity for some people who eat lots of yogurt, but we do not know if they are true.

Gluconic acid improves the life span of fruit flies slightly more than lactic acid does, but I cannot think of an economical and convenient source of this molecule. It is an approved food additive and of low cost, but used in few products. Gluconic acid is produced by the fermentation of glucose, which is the sugar in the blood. Both lactic and gluconic acids are the products of sugar fermentation. While these two aldonic acids considerably increase the life span of Drosophila, their affect on the life span and general health of humans is unknown.

Chapter 22

ANTI-INFLAMMATORY AGENTS

The use of anti-inflammatory agents, such as aspirin, is widespread. Many individuals consume large amounts for prolonged periods of time. How these compounds affect life span would seem to be of general interest.

Years ago it was reported by physicians in California that people who consumed aspirin on a daily basis seldom had heart attacks. This was a very important observation and has led to the use of aspirin to prevent heart attacks and strokes by millions of people. Since heart disease is an age-related condition, it seems probable that aspirin use and aging might be somehow connected. The use of anti-inflammatory compounds such as ibuprofen also has been reported to greatly reduce the incidence of Alzheimer's disease.

Hochschild (Effect of Membrane Stabilizing Drugs on Mortality in Drosophila melanogaster, Experimental Gerontology, vol. 6, pp. 133-151, 1971) showed in a series of experiments that certain anti-inflammatory compounds could improve the life span of fruit flies. With aspirin, for example, he found a 13 to 40% increase in life span. He proposed that stabilization of lysosomal membranes was the reason for greater survival. Lysosomes are small organelles in the cell which release enzymes to remove cellular debris. They should not be active unless there is a reason. Their enzymes could degrade important normal molecules if not kept in check. The theory is that aspirin and similar compounds prevent the release of lysosomal enzymes.

Two years later, however, he reported in a series of experiments with female C57BL/6J mice that he was unable to find a significant improvement in life span with anti-inflammatory agents. This was, of course, a great disappointment to him and everyone else. However, it is notable that he found no decrease in life span.

Fruit flies and pain relievers

We examined the effect of a number of anti-inflammatory agents on the life span of fruit flies. The results were disappointing. Many of them reduced life span at high concentrations. Only 2,3-dihydroxybenzoic acid improved life span by 10.9%. Acetylsalicylic acid, which is aspirin, reduced or gave no improvement in life span. We found that Bufferin failed to improve life span and reduced it at high doses. Other lesser-known anti-inflammatory agents such as chlorpromazine, dimethysulfoxide, indomethacin, phenol, salicylic acid and sodium salicylate all failed to prolong life span and reduced life span considerably at high doses.

It is of interest that anti-inflammatory agents inhibit the production of hydroxyl free radicals to a considerable extent (Sagone, A.L. Jr., In Rodgers and Powers (Eds.), Oxygen and Oxy-radicals in Chemistry and Biology, Academic Press, New York, 1981). Sagone found that phenol and 2,3-dihydroxybenzoic acid (this is the only compound which we found to prolong life span) inhibited hydroxyl radical production by 90 and 94%, respectively. Indomethacin, aspirin and dimethysulfoxide also inhibited hydroxyl radical production but to a lesser extent. It is not clear whether anti-inflammatories react directly with hydroxyl radicals or proceed by an indirect mechanism. It is clear, however, that none of the anti-inflammatory agents decreased life span when given at the low concentrations that most humans would be exposed to. It can be concluded that continuous exposure to these valuable medicinal agents appears to represent little or no risk to Drosophila in terms of changes in life span. Whether or not these observations can be extended to humans remains to be determined.

The whole question of anti-inflammatory agents and the aging process is of serious importance although given little attention. I have not seen any studies of life span changes with the newer anti-inflammatory agents, such as the Cox-2 inhibitors, that do not cause stomach acid problems. One of these, Vioxx, has recently been withdrawn from the market because it may cause heart attacks and stroke. The reason for this seems to be that Cox-2 inhibitors block the production of prostacyclin that is necessary for keeping blood vessels from constricting. Constricted blood vessels can sometimes lead to a stroke or heart attack.

Reading the list of inactive ingredients added to most anti-inflammatory products can be interesting. For instance, adding the non-digestible sugar, sorbitol, to a pill that you do not chew and which causes intestinal distress in many people seems unnecessary. Perhaps the worst of all is adding iron oxide in order to give that pill a nice color. I went to the drug store yesterday just to look at the number of pain relievers with iron oxide as the major inactive ingredient. It was many of them.

Chapter 23

VITAMIN A

V itamin A was first shown by Sherman et al. (Proceeding of the National Academy of Sciences, vol. 31, pp. 107-109, 1945), and by Paul and Paul (Journal of Nutrition, vol. 31, pp. 67-78, 1946) to influence the length of life of rats. The life span increased as the dietary intake of vitamin A was increased from 3 to 12 international units/gram of food in the Sherman et al. study. The life extension was 18.8% for female rats at 12.0 I.U./gram of food and 10.8% for males when compared to a lower dose of 3.0 I.U/gram of food (I.U. is an abbreviation for international unit). Paul and Paul reported a similar result for both male and female rats for vitamin A given at concentrations ranging from 1 to 20 I.U./100 grams of body weight. The changes in life span in the Paul and Paul study were enormous. In one group getting only 1.0 I.U. of vitamin A /100 grams of body weight, the life span was very short. However, increasing the amount of vitamin A in the diet to 20.0 I.U./100 grams of body weight improved life span by 754% when compared to the group getting only 1.0 I.U./100 grams of body weight. Clearly, a low amount of dietary vitamin A greatly reduced life span, whereas increasing it resulted in a dramatic increase in life span.

In a later study, however, Sherman and Trupp (Journal of Nutrition, vol. 37, pp. 467-474, 1949) found that increasing the vitamin A content of the diet from 12.0 to 24.0 I.U./gm of food actually reduced the average length of life for rats, 9.6% for males, but only 2.1% for females. Apparently males are more sensitive to vitamin A toxicity. This ability of vitamin A to both prolong and reduce life span has not been reported for other organisms, nor has it been adequately explained.

In agreement with these early studies, we found that different concentrations of vitamin A both prolong and greatly reduce the life span of Drosophila, depending upon the dietary content of vitamin A. We also found a possible explanation for this phenomenon based on our observation that vitamin A compounds react with superoxide radicals. These are the free radicals which can cause considerable biological damage.

The well known early symptoms of vitamin A deficiency include night blindness, swollen and red eyes, increased susceptibility to infection, abnormal skin cells characterized as "frog skin" and lack of skeletal growth and bone abnormalities in the young.

Although the metabolic function of vitamin A is not fully resolved, there is ample documentation of its biochemical deficiency symptoms. Thus, low vitamin A levels have been found to result in decreased activity of gulonolactone oxidase, codeine demethylase, squalene cyclodehydrase, ATP-sulfurylase and sulfate transferase. These are all important enzymes and indicate that something is wrong. Furthermore, protein synthesis by membrane-bound polyribosomes has been shown to be decreased under conditions of vitamin A deficiency. All of this is biochemical evidence that vitamin A deficiency is a problem.

It has been suggested that vitamin A might function as an antioxidant because of its five double bonds (bonds which can readily react with free radicals). Consistent with this suggestion is the observation that vitamin A seems to act as a strong inhibitor of lipid peroxidation. We found that vitamin A compounds at low concentrations inhibit lipid peroxidation, but, surprisingly, at high concentrations they stimulated the peroxidation of lipids such as linolenic acid.

Fruit flies and vitamin A
When flies were reared and maintained as adults, on medium containing additional dietary Vitamin A, the median life span was increased by 17.5% at a vitamin A concentration of 2.5 I.U./gm of food. The entire survival curve was shifted to greater survival times when compared to the normal control diet containing vitamin A at a concentration of 0.374 I.U./gm of food. A similar increase in life span was observed at a vitamin A concentration of 25.0 I.U./gm of food. At higher concentrations of vitamin A, however, reduced survival times were observed. Extrapolation of the data for both increased and decreased survival times indicated an optimum vitamin A concentration of 8.0 I.U./gm of food for maximum survival.

However, when flies were placed as newly emerged adults on medium containing additional vitamin A, there was no improvement in life span. For example, vitamin A concentrations of 0.25, 2.5 and 25.0 I.U./gm of food produced no significant change in the median life span when given during the entire adult stage. The conclusion is that vitamin A must be incorporated during the developmental stages in order to have any positive influence on the adult life span of Drosophila. It is not known if this observation extends to other organisms, including humans.

Vitamin A concentrations have been reported to change with aging in both mice and humans. In general, very high concentrations are found in the liver of both older mice and older humans. Whether or not these large age-related increases represent a risk remains to be seen. Excess vitamin A reduces the life span of both rats and Drosophila. It is also remarkable that the dietary concentrations of vitamin A required for influencing the length of life are so similar for both rats and fruit flies, with 12.0 and 8.0 I.U./gm of food being the optimum concentration for rats and flies, respectively.

The observation that vitamin A influences the length of adult life span in fruit flies when given during the developmental stages is consistent with reports of gene activation by vitamin A compounds. Genetic determinants of longevity may be regulated by the activation of receptors by dietary vitamin A at certain critical times during growth and development. This is why it is important for children to have adequate dietary vitamin A. Without adequate amounts they can, and do, go permanently blind, among other serious problems.

Excess vitamin A and aging

However, excess vitamin A exposure during the adult phase of life may counteract the beneficial influence of exposure during the developmental stages. It has long been known that excess vitamin A is toxic. At present, the optimum concentration of dietary vitamin A during the adult stage of life is unknown, except for fruit flies. Processes other than aging itself may be affected differentially by low and high dietary intakes of vitamin A. For example, it is known that decreased immune response occurs at low, but also at high, dietary intakes of vitamin A.

Mitochondrial membranes appear to be the primary target of excess vitamin A exposure, but they are also a target during vitamin A deficiency.

The cytotoxicity of vitamin A for both yeast and human cells can be overcome to a large extent by the presence of antioxidants such as ascorbic acid, vitamin E and glutathione. This is consistent with the free radical mechanism of action that we found. Therefore, vitamin A can function as an antioxidant at low concentrations but as an oxidant at high concentrations. Other molecules have been observed to do the same, including the enzyme superoxide dismutase (SOD).

The maximum life span reported in the Sherman (1945) studies was achieved at 12.0 I.U./gram of food or 3.2 I.U./calorie of food. For humans this would amount to 9,600 I.U./day for a man consuming 3,000 calories/day. The minimum recommended daily requirement is 5,000 I.U./day for adult humans according to the National Research Council which would be about right for a person consuming 1,500 calories/day based on the rat studies. Does this mean that we are not getting enough vitamin A in the diet?

The real question is: what is the requirement for vitamin A in adult organisms including humans? There is no doubt that a small amount is required for normal development, but the level required for adult nutrition is unknown. Since it accumulates in the liver with aging in both rodents and humans to very high levels, this may represent a health hazard in terms of the aging process. Unfortunately, this subject has not been studied. In humans, the liver content of vitamin A increases with every decade of life for most, but not all. This is not true for alcoholics and people suffering deficient diets or starvation. Excessive alcohol consumption reduces the vitamin A content of the human liver by as much as 86% and eventually lowers that in the blood. Starvation lowers the level of almost all nutrients, including vitamin A.

The truly disturbing fact about vitamin A in the diet is that many of the problems of deficiency are also seen when an excessive amount is consumed. The most prominent of these problems is that of bone growth and development. Low dietary levels of vitamin A result in lack of bone growth and fragility with high amounts of calcium excreted in the urine. At high amounts of vitamin A in the diet, the same thing occurs with bone loss, fragile bones, and pain in the joints and bones. A general feebleness occurs at low levels of dietary vitamin A, and yet we see the same thing at high dietary intakes in humans. High doses of dietary vitamin A cause the mitochondria, which are the energy producing organelles of the cells, to swell, but low doses of dietary vitamin A do the same thing. Much, if not all, of this can be explained by a long scientific discussion, which will not be presented here, where superoxide and hydroxyl free radicals are at the basis of the problem.

Alcoholics are well known to develop cirrhosis of the liver, and this could be due to the vitamin A deficiency seen in most alcoholics. Excess vitamin A consumption, incredibly, also does the same thing. In one report, an individual, who did not drink alcohol, who was taking 25,000 I.U./day of vitamin A for a period of 6 years developed cirrhosis of the liver. Others, who drank very little or no alcohol but who were taking 100,000 I.U./day for an average time of 9.3 years, developed the same liver problem, and 22% of them died in 4.5 years over the period of the study, even though they had stopped taking vitamin

A supplements (Geubel, De Galocsy, Alves, Rahier and Dive, Gastroenterology, vol. 100, pp. 1701-1709, 1991).

Thus, we know that 25,000 I.U./day is toxic to humans, and that 100,000 I.U./day can result in death at an early age from cirrhosis of the liver. The toxic dose is only 5 times the recommended daily dose of 5,000 I.U./day for an adult human. A dose of 100,000 I.U./day over a period of 9.3 years was a very serious problem, even after these people stopped taking vitamin A supplements. This is because vitamin A is stored in the liver and not excreted. In terms of the problem of human aging, this is a very important observation, especially since cessation of the vitamin A supplements did not cure the problem.

Since the exact human requirement for vitamin A is not known, especially for older adults, we can only guess what it might be. Deficiency symptoms include retarded growth, night blindness, swelling of the eyelids with mucus secretion and, finally, blindness, thickening of the skin, dry and itchy skin, additional degenerative changes in the eyes, poor bone and tooth development, a very high incidence of several forms of cancer and stones or sand in the form of calcium phosphate in the urine. Numerous studies in animals have shown that vitamin A deficiency results in the appearance of many kinds of cancer which can be prevented with an adequate dietary intake of vitamin A. This is of enormous importance in terms of cancer control and prevention. If these problems occur, then increased dietary vitamin A is probably needed.

Excess vitamin A and illness
Surprisingly, the effects of vitamin A overdose in humans and animals are equally serious. These include irritability, headaches, dizziness, nausea, vomiting, loss of appetite, nerve lesions, fatigue, feebleness, difficulty in sleeping,

painful bones and joints, tenderness in the soft tissues above the joints, spontaneous fractures of the long bones, bulging eyes, excess mucous formation, abnormal bone growth and brittle bones, loss of hair, brittle finger nails, jaundice and cirrhosis of the liver, dry and itchy skin, peeling skin with reddening and itching, inflammation of the gums, weight loss, calcification of the arteries and the heart and decreased clotting time of the blood. Vitamin A excess can also cause cardiovascular disease and massive bone loss. Please see: "Cardiovascular Calcification in Rats with Hypervitaminosis A," by Strebel, Girerd and Wagner, Arch. Path. vol. 87, pp. 290-297, 1969. The doses of vitamin A used in this study with rats, however, were very high (50,000 I.U./day for 16 days).

Excessive intake of vitamin A can even be a problem for the very young. Since excess vitamin A is known to produce birth defects in animals and humans, it has been recommended that pregnant women limit their total daily intake to no more than 8,000 I.U./day. It should be noted that large doses of vitamin A on a short-term basis are needed to produce all of the above problems in both humans and laboratory animals. The important question is: can slightly higher than normal doses do the same things over a period of many decades of human life?

The symptoms of vitamin A excess sort of sound like aging to me. Is it possible that vitamin A poisoning is a major cause of human aging? This is, of course, a bold assertion, but we can review the evidence since this whole book is really a detective story of sorts. The maximum life span in the Paul and Paul (1946) study with rats was achieved at a vitamin A concentration of 20 I.U./100grams of body weight, which would translate to 14,000 I.U./day for an average male human weighing 70 kilograms. This is a higher amount than that found in the Sherman study with rats noted above and is also very close to the 25,000 I.U./day value known to produce liver disease in a human when consumed over a period of 6.1 years. The issue of a much higher metabolic rate in rats compared to humans has, also, not been adequately addressed.

Almost all nutritional studies have been done on young and growing organisms. The lack of vitamin A, or any other nutrient for that matter, can produce very serious consequences. Practically no studies have been done on the nutritional needs or lack of needs in adult or older organisms using animal models. The research on humans is also limited. At this point, one just has to make guesses. Vitamin A is like iron; there is no way to get excessive amounts out of the body. With iron you can donate blood to reduce the body content of iron, but with vitamin A the only thing you can do is stop taking vitamin supplements or change to a vegetarian diet (since there is no vitamin A in plant foods; they contain only beta-carotene which can be converted to vitamin A). These changes, however, might increase your chances of developing cancer since adequate dietary vitamin A is known to decrease the incidence of tumors.

The fact is that we simply do not know what the requirement of vitamin A is for adult humans, especially for older humans. The one study I know of measuring vitamin A in the blood of people of different ages shows an enormous amount of variation with no apparent trend. Blood level determinations of vitamin A are known to not always reflect the total body content of vitamin A. No one, however, wants to have a liver biopsy, which would be the best way to measure the body content of vitamin A.

The reason I believe that vitamin A toxicity may be involved in the aging process is how the weight versus age curves change in mice and rats. Just prior to death, there is always a weight loss in these rodent colonies in every study that I have ever seen. As the death rate increases so does the degree of weight loss. Excessive vitamin A intake in young animals and humans also causes a considerable weight loss within a short period of time. Since vitamin A accumulates with age in both human and rodent livers on what we consider to be "adequate" diets, it is possible that this accumulation causes the gradual and prolonged weight loss during normal aging in mice and rats. We do not know if this occurs in humans, since it seems that there are no studies related to this observation.

Vitamin A deficiency

While excessive dietary intake of vitamin A may be a worldwide problem for many adults, it is not the case for millions of other people, especially children. I have tried to find what the first sign of vitamin A deficiency is in humans, and the best that I can come up with is from an old textbook of biochemistry (West and Todd, Textbook of Biochemistry, Macmillan, New York, 1951, p. 706):

"The skin lesions vary considerably in different individuals. The general features of the vitamin A deficiency involve dryness and roughness of the skin. This appears early in the deficiency and is due to suppression of the sweat glands. A keratosis (horny growth), especially of the hair follicles, is a prominent feature. Papules (red elevated areas on the skin), masses of keratinized epithelium (harden skin) develop, and these are readily felt by rubbing a finger over the involved area. The sides and backs of the thighs and the lateral parts of the forearm are most frequently involved, although the condition is often more extensive. Such lesions are considered by many to be one of the earliest symptoms of vitamin A deficiency in humans. Some reports concern the presence of the skin condition without perceptible abnormalities of the eyes, indicating that the former condition may precede the latter."

I had these symptoms for a while as a child, which is the time when most symptoms of vitamin A deficiency appear, and so did my wife. In her neighborhood almost all the kids had it, and they called it "chicken skin."

Vitamin A excess

The "chicken skin" test seems to me to be a good way to monitor whether or not children or adults have even a marginal vitamin A deficiency, but what about the problem of vitamin A toxicity? How can a person tell early on if they are consuming too much vitamin A? Unfortunately there is no simple test. If anyone wishes to pursue this problem, I recommend the article by Hans Biesalski in Toxicololgy, vol. 57, pp. 117-161, 1989. His extensive review of the vitamin A literature indicates that perhaps the first essential feature of excess vitamin A is peeling and reddening with itching of the skin and pain in the bones.

As a result of writing this chapter, I looked at what my normal vitamin A intake would be without taking any vitamin pills. I eat an egg (400 I.U.) in the morning with butter (100 I.U.) on my toast and drink a cup of buttermilk (500 I.U.) at lunchtime. The total vitamin A intake for my day is 5 times less than the 5,000 I.U./day recommended daily allowance (RDA). Perhaps I should be taking a vitamin pill. Probably not, since most of the vitamin A in my diet comes from the conversion of carotenoids to vitamin A in the intestine.

One of these carotenoids is beta-carotene. This is the compound which gives carrots their characteristic orange color. There is a great deal of beta-carotene in not only carrots, but also in other vegetables, especially leafy greens such as lettuce and spinach. In the intestine each molecule of beta-carotene is split to form two molecules of vitamin A, but only if you need vitamin A. Beta-carotene, therefore, does not produce vitamin A toxicity as pure vitamin A can. Just 6 leaves of romaine lettuce, for example, contain enough beta-carotene to produce 1,000 I.U. of vitamin A. The average balanced diet, therefore, probably provides all the vitamin A needed by the body, but the fact is that we do not know what the optimum dietary intake of vitamin A is for adult humans over the period of their entire life span. This problem could probably be resolved with a comparatively small amount of research.

Meanwhile, every person has to weigh the evidence and decide for himself or herself, based on his or her personal life style, what an acceptable dietary intake of vitamin A should be. It should be stressed, however, that excessive vitamin A intake can be very toxic. The recommended daily dose of vitamin A has recently been reduced to 3,000 I.U. for men and to 2,300 I.U. for women based on new evidence that higher intakes of vitamin A cause osteoporosis in humans. There are still many unanswered questions remaining to be investigated.

Chapter 24

MERCURY

Mercury in the form of metallic, organic and inorganic mercury compounds is widely used in industry and commerce. "Silver" dental fillings, for example, contain up to 50% of pure elemental mercury in the amalgam used for filling teeth. Acute exposure to mercury from contaminated food consumption and occupational exposure has resulted in serious illness and even death. High-level mercury pollution of the environment has been, and continues to be, a major public health concern. The effect of continuous low-level exposure, however, and the question of whether or not mercury exposure affects the aging process have not been adequately addressed.

Fish, mice and fruit flies
Bache, et al. (Science, vol. 172, pp. 951-952, 1971) reported that the total mercury content of trout from Cayuga Lake in New York State increased with aging. On the other hand, Westoo (Science, vol. 81, pp. 567-568, 1973) found no change with age in the mercury content of salmon and sea trout of various ages from Swedish rivers. This was in agreement with his previous findings for pike of different

ages caught in Swedish waters. It is not clear whether or not the disagreement in these reports is due to differences in mercury exposure, species, organs examined or analytical technique.

In view of the serious neurological damage known to result from mercury exposure and its possible involvement in the aging process, we examined whether or not mercury accumulates with age in another species, the mouse. We addressed the question of the extent of mercury accumulation with age in mice maintained under normal care conditions in a conventional rodent colony without exposure to known mercury sources other than those normally found present in food, water and air. We also measured the effect of various concentrations of added dietary mercury on the life span of the fruit fly.

The amount of mercury in the sixty-five mouse organs which we examined ranged from the lowest value of 0.0051 ppm (parts per million) found in the liver of a young animal to the highest value of 1.14 ppm found in the brain for adult mice ranging in age from 133-904 days. We found that most of the mercury accumulation occurs in the brain, 200 times more than in the liver. The concentration of mercury increased with age in the brain, kidney and heart, and declined with age in the lung and liver. These changes, however, were not statistically significant due to scatter in the data.

Mercury in the brain

Nevertheless, it is clear that a considerable increase in the concentration of mercury in the brain was found in some, but not all, older animals. This suggested that perhaps some other variable was involved. Since mercury is excreted primarily by the kidneys through the urine and by the liver through the bile, we calculated the brain to kidney and brain to liver ratios for mercury content from the data. The result was a significant increase with aging in the brain/kidney ratio of total mercury content. An even more significant relationship was found for the brain to liver ratio of total mercury content. The increase in this ratio with age was exponential. Again, it is important to note that aging itself is an exponential process with the rate of aging in humans doubling every seven years.

Apparently, the ability to maintain a low brain to liver ratio of total mercury content decreases with age regardless of the mercury content of the brain or liver of individual mice. Thus, while some older mice appear to be able to maintain low concentrations of mercury in the brain, they are unable to achieve the low brain to liver mercury ratios seen in younger mice. This also seemed to be the case for the brain to kidney ratio.

We also found a significant relationship been the lung to liver mercury ratio and aging. This result suggests that the lung may represent a possible absorption site for mercury, most likely from the air. All other ratios such as brain to lung, brain to heart and heart to liver failed to show any significant relationship with age of the mice.

Fruit flies

We examined the effect of dietary mercury on the rate of aging on fruit flies. Three different concentrations of mercury were added to the food to give final concentrations of 200 ppm, 2,000 ppm and 20,000 ppm of mercury in the diet. Incidentally, these high concentrations of mercury would certainly kill a mouse or a human. Only the highest concentration significantly reduced the median life span by 4.3%. The age at which 80% of the flies remained alive and the age at which only 20% remained alive were also both reduced at this highest level of mercury. Surprisingly, the lowest concentration of mercury, 200 ppm, actually increased the median survival time, but the increase was not significant. Overall, increased dietary mercury had little or no influence on the rate of aging in fruit flies, as reflected by survival times.

Mercury toxicity, disease, and dementia in humans

The principal result of our investigation was that the total mercury content of the organs from mice maintained under conventional environmental conditions does not significantly increase with aging when the variability in the data is considered. This is in agreement with the previous report of Westoo (1973) for fish. It should be noted, however, that most of the fish analyzed by Westoo were from one to two years of age with the oldest being seven years of age. Bache et al. (1971), on the other hand, examined a wider range of ages of lake trout (between 1 and 12 years of age) and found higher concentrations of total mercury (up to 0.7 ppm) in their fish. This suggests that possibly the environmental exposure levels of mercury may have been higher in the Bache et al. study. One should also consider that they looked at much older fish. Environmental exposure levels and age may, therefore, determine whether or not an age-related increase in mercury is found in various organisms. The effect of different levels of environmental mercury exposure on the aging process is, however, unclear. The questions of whether or not mercury accumulation occurs with aging in humans, and how it affects life span have not been studied.

There is no doubt, however, that exposure to elevated concentrations of mercury in humans from either dietary or atmospheric sources can lead to serious neurological conditions including paralysis, deafness, blindness, spasticity and seizures. Volatile organic mercury compounds which used to be used in latex paints, for example, can represent a considerable risk. The amount of mercury in the environment depends on numerous factors, both natural and man-made. The burning of coal, for example, adds 3,000 tons of mercury per year to the atmosphere.

Whether or not this and other sources of environmental mercury contribute to subclinical neurological changes in man and other organisms remains to be seen. It is of interest to note, however, that mercury-containing reagents such as p-chloromecuribenzoate inactivate the inhibitor of the enzyme, alkaline ribonuclease, in the brain. This enzyme breaks down ribonucleic acid (RNA), resulting in lower levels of RNA. The inhibitor binds to the enzyme to prevent this from happening. Mercury compounds in the cell inactivate this inhibitor. It is of considerable interest that decreased total RNA, and increased free ribonuclease as a result of lower inhibitor-bound ribonuclease levels, have been reported in individuals with Alzheimer's disease.

It has also been demonstrated that the brains from Alzheimer's patients contain almost twice as much mercury as those from age-matched control patients. This is a startling result since it is known that the primary problem with mercury exposure is neurological damage. Our results with mice offer a possible explanation for this observation. The increase with aging, in the ratio of the amount of mercury in the brain relative to that in the organs representing major excretion sites for mercury, suggests that mercury removal from the brain is less efficient in older organisms. This decreased efficiency could lead to elevated mercury concentrations in the brain for some, but not all, individuals, as appears to be the case for our results for mouse brain.

Thus, while mercury does not appear to be an initiator

of the aging process in Drosophila, it may be of significance in the development of human aging-related diseases, such as Alzheimer's disease. It would be of interest to know, for example, if individuals exposed to higher than normal amounts of mercury during the later part of their life span are more likely to develop neurological disorders. The presence of "silver" dental fillings might represent one possible source of mercury exposure which could affect the elderly, but not the young, in a negative manner due to the reduced excretion capacity of the liver and kidney. Other sources of mercury exposure are more likely, especially alkaline batteries, paint, broken thermometers, electronics, contaminated air, broken florescent light bulbs (they contain up to 1 gram of mercury) and food like tuna fish. These are issues that have been pretty much ignored.

One of the most striking results of our studies with fruit flies and mercury was their incredible tolerance for high levels of dietary mercury, levels that would kill most animals and humans. The same was true of their high tolerance of dietary lead. There must be something special about fruit flies and their amazing tolerance to toxic metal ions. Just as a proposal, suppose that mercury intoxication is a reason for Alzheimer's disease. Fruit flies can deal with this, but humans cannot. It seems that this might be a reason for giving fruit flies an evaluation in terms of this disease state.

Chapter 25

SPECULATIONS

The research reported here involved projects that were mostly carried out with the collaboration and help of many very talented and capable people. Their names are given in the references listed at the back of the book. There are many other areas of aging research that I have not dealt with. Therefore, this book is far from being the source of all information concerning the aging process. One of the studies I have not been involved with is "dietary restriction." Very large increases in life span, on the order of 50% and more, have been achieved by simply limiting the food intake of rats and mice at a young age. The reason for this effect is unknown, and it does not apply to all strains of mice. For example, the life span of C57BL/6J mice has not been improved by dietary restriction. This is the strain that we used for all of our mouse experiments. I am, therefore, really not qualified to make comments on this most interesting phenomenon. There are other areas of aging research that are equally important, but, again, I feel that it is better for those directly involved to tell the story.

The use of fruit flies, or any other model system for that

matter, to study the human aging process is flawed. The results with mercury and lead in fruit flies clearly point to this fact. Even at levels which would kill any human, fruit flies experience no negative effects in terms of life span changes. We must, therefore, consider changes in the life span of fruit flies, mice, rats and other organisms to be only possible indicators of what might occur in humans.

The only thing which can be said with confidence is that very large changes in the rate of aging of fruit flies and other small organisms can be achieved in the laboratory by changing temperature, ultraviolet and visible light, radiation, oxygen, diet, toxic metal ions, exercise and chemical additions. Extending these results to humans requires putting a lot of pieces of the puzzle together and making careful observations of the human response to a wide variety of factors. It is known, for example, that studying atherosclerosis in mice and rats is very difficult. On the other hand, rabbits, pigeons and other animals readily develop this disease when fed cholesterol or lead, as well as by other means. Every model system has its limitations, but without these different systems, biomedical research would not be possible.

Almost all vitamin research was originally done with mice and rats. The first vitamin to be discovered in rats was vitamin A, and all the others followed, with the exception of vitamin C. Rats and mice do not need vitamin C, since they make it themselves. Humans, on the other hand, must have this vitamin in their diet. Therefore, the rodent model system for studying essential vitamins becomes flawed when we consider the need for vitamin C in humans. This is another example of the need to evaluate the limitations of any model system.

Considering this, one could ask an interesting question. Have all the vitamins needed by humans been discovered?

If humans are considered to be unique, then the answer is probably no. They have not yet been found. Among all the vitamins presently known to be required by man, two are antioxidants, vitamin E and vitamin C. They function primarily by protecting other molecules from the damaging effects of oxygen. Vitamins are small molecules that are required at relatively low concentrations in order for life to continue. Without them the organism dies, as happens when humans develop scurvy in the absence of dietary vitamin C. It is possible that human aging, as we know it today, could be the result of an, as yet unknown, vitamin deficiency, especially an unknown antioxidant.

Within the plant kingdom, there are literally thousands of possible molecules which might answer this question. We have yet to explore them. One of these molecules is lycopene. It occurs in watermelon, pink grapefruit and tomatoes. Much epidemiological evidence indicates that people who consume tomatoes suffer from less cancer and cardiovascular disease. Perhaps, it also decreases the rate of aging based on its antioxidant properties. One could even speculate that lycopene is a new vitamin. The vitamin companies are already adding it to their pills, even though we are not really sure of the consequences.

When everything is considered, there are four basic possibilities for the cause of human aging.

(1) First is that we lack some important compound or group of compounds in our diet that could be referred to as unknown vitamins. This could result in a slow death which we now label as aging.

(2) Second is that we are slowly being poisoned by radiation (and this includes light), and by atoms and molecules in our diet and the environment. Cadmium is the most likely candidate for this process, followed by iron,

copper, lead, aluminum and mercury. Even excess vitamin A accumulation is a possibility.

(3) Third is that there is a biological clock that begins to tick when the egg of your existence in your mother is first formed. This clock could be the original mitochondria which comes entirely from your mother. In this model, aging begins even before conception and, when the mitochondria are finally damaged beyond repair, your life ends. Since the DNA in the mitochondria, for some strange reason, cannot be repaired, this could be the molecular basis of the biological clock. If this is true, there is not a whole lot that can be done about it, unless something like cadmium toxicity is the reason for the shut down of the biological clock in mitochondria.

(4) Fourth is that aging is simply due to bad luck, high-risk behavior and poor medical care. The list of possibilities here could be very long: dangerous jobs, stressful jobs, excess alcohol or smoking, malnutrition, excess nutrition, infection, depression, little exercise or too much exercise, aggressive behavior, criminal assault, drug addiction, sex related diseases, mental disease, vehicular accidents, poor hygiene, job loss, divorce, money stress problems, fires, death in the family, contaminated water or air, war, natural disasters and just plain wear and tear on the body.

My personal pick is number one, although number four is close to the mark. Probably all four reasons for aging are valid which makes the story more complicated. It would help if we knew what the real maximum human life span is. Several gerontologists have said that it is 120 years, based on the observation that almost no one has lived to be older in modern times. However, there are many reports of much greater life span throughout human history. Claims of 150, 180, 300 and even 969 years of age have been made. In March 2004, the world's oldest documented man, Joan Riudavets-Moll, died on Spain's Menorca Island. He was active and articulate up until the day before he died. He smoked, "not too much," but attributed his long life to moderation, exercise, a healthy diet and a calm life. He was 114 years old. The oldest living woman at the time of this writing is Charlotte Benkner of Ohio, U.S.A. who is 115.

How did these people achieve such a long life span? The life insurance companies have collected information on this subject and have come up with a few interesting observations. Women live 5.3 years longer than men (it used to be more than this), married people live 7 years longer than single people, people who live in the country live 2 years longer than those living in a city, individuals with a religious faith live 7 years longer than atheists, being overweight can greatly reduce your life span by as much as 10 to 20 years or more (especially if you have diabetes), and cigarette smoking reduces life span by as much as 7 years or more, if one smokes a lot. If you apply for life insurance most companies consider these things. These factors are of interest, but, really, what can one do about them? Not much, unless people are willing to make large changes in their life styles. These are, of course, generalities and may or may not apply to any given individual. None of these changes, however, is going to get anyone up to the 114 years of age of Joan Riudavets-Moll of Spain, who enjoyed a full and active life until one day before his death.

Why did Joan live so long, and why was he in such good health? Is this a possibility for the rest of us? Perhaps it is, with the application of what we already know about the basic aging process in model systems of other organisms. Every person has to view the scientific material presented here and elsewhere with skepticism and wonder if it has relevance to humans. Some, and probably most, of it does apply to humans, but it could be a long time before it is confirmed. If most of it does apply to humans, then one would have to conclude that Joan died at a relatively young age. The fact is that aging research is a much ignored and greatly underfunded effort. Only the general public can change this situation.

GLOSSARY

Activation Energy, energy or heat needed to get a chemical reaction to start

Age Pigments, small particles which accumulate with age in most cells, especially in the heart

Aging, an undefined process that leads to a greater probability of death from all causes, including disease and accidents

Aldonic acid, acid of a sugar, examples are lactic and gluconic acids

Alzheimer's Disease, a form of dementia of unknown cause in older people

Antioxidant, molecule which reacts with oxygen or oxygen containing molecules

Antioxidant Enzyme, protein that removes oxygen-containing molecules

Background Radiation, small amount of natural environmental radiation

Barn, unit for measuring the ability to capture neutron radiation

Beta-carotene, a carotenoid found in plants, an antioxidant, reacts with singlet oxygen, can be converted to vitamin A

BHA, synthetic antioxidant commonly used in many food products

BHT, synthetic antioxidant used in food products

Cadmium, very toxic metal with no known biochemical function

Carcinogens, chemicals, both natural and synthetic, that cause cancer

Catalase, antioxidant enzyme that removes hydrogen peroxide from the cells

Catalyst, chemical or enzyme that greatly increases the rate of a chemical reaction

Celsius, metric system measure of temperature used by most scientists, 30 degrees Celsius equals 86 degrees Fahrenheit, for example

Ceroid, an age pigment usually found only in disease states

Ceruloplasmin, copper-transporting protein found in the blood

Chelator, chemical that wraps around metal ions ("chela" is Greek for claw)

Chromium, essential trace element, can be beneficial or toxic depending on electronic charge

Chromosomes, structures in the nucleus that contain genetic material, the DNA

Collagen, structural protein making up most of the material in joints, bone and other organs

Collagen Theory of Aging, collagen becomes old and brittle with time and this is possibly why we age

Cosmic Rays, radiation from outer space

Cysteine, amino acid containing sulfur, functions as an antioxidant

Denham Harman, father of the free radical theory of aging

DABCO, water-soluble compound that reacts with singlet oxygen

Boron, essential nutrient but can be toxic, seems to prevent osteoporosis

DNA, deoxyribonucleic acid, the genetic material found in the nucleus of all cells

Drosophila, fruit fly insects

EDTA, commonly used metal ion chelator in food products, also used by physicians to remove lead from the body

Electron, small subatomic particle with a negative charge spinning around the nucleus or center of an atom

Enzyme, a protein that carries out a specific task such as removing free radicals

Ferritin, iron-transporting protein

Free Radical, any molecule or atom with an unpaired or free electron

Fluorescence, emission of visible light when molecules are exposed to ultraviolet light

Fluorescent Granules, fluorescent age pigments, composed of fat, protein and metal ions, found in all older organisms including humans

Glucose, sugar found in blood

Glutathione, a tripeptide meaning it is composed of three amino acids, needed for the action of the enzyme glutathione peroxidase

Glutathione Peroxidase, antioxidant enzyme, protects fats and oils by removing toxic peroxides

Glycoprotein, protein containing both protein and sugar

GTF, glucose tolerant factor, contains chromium and is

believed to lower blood sugar levels and the incidence of diabetes

Hormesis, favorable response to radiation and other environment toxicants

Hydroxyl Radical, most damaging of all the free radicals, composed of one hydrogen atom and one oxygen atom with an extra electron

Interleukin-2, immune system molecule

I.U.. International Unit, used to express the amount of vitamin A

Lactose, sugar found in milk, can be converted to lactic acid by fermentation of certain bacteria

Lead, widely used metal with no known biological function, can be very toxic to humans even at low concentrations

Lipid, molecule of fat

Lipid Peroxidation, reaction of oxygen with fats and oils resulting in the production of peroxides and free radicals

Lipid Peroxidation Theory of Aging, biological organisms go rancid with time and this is the reason for aging

Lipofuscin, age pigment, glows in the dark, also referred to as cellular garbage

Lux, a measure of light exposure

Lysosome, small particle or organelle in the cell that removes damaged debris from the cell

Malonyldialdehyde, chemical product formed during the lipid peroxidation process

Maximum Life Span, age at which the last organism in a given study dies

Mean Life Span, average life span

Median Life Span, age at which half of the organisms are alive and half dead

Membrane, structure surrounding the cell and other cellular particles

Mercury, only metal to be a liquid, can be very toxic, no known biological function

Metabolic Rate, amount of heat or energy production per unit of time

Metal Ion, ionic form of a metal rather than the pure metal, almost all metals form ionic salts such as sodium chloride, common table salt

Metallothionein, protein that protects against toxic metal ions such as cadmium

Methionine, essential amino acid containing sulfur, also is an antioxidant

Methuselah, longest living human in recorded history, 969 years

Microsome, small organelle in the cell that removes toxic chemicals

Mitochondria, small particles or organelles in the cell that produce energy and heat

Mitochondrion, the singular of mitochondria

Mitochondrial DNA, the genetic material inside the mitochondrial particle

Mutagenic, causes DNA damage

NDGA, natural food antioxidant formerly used to preserve fats and oils

Neutron, dangerous radioactive particle found at low levels in the environment

Nitrogen Dioxide, smog pollutant, also can initiate lipid peroxidation

Nucleolus, a compartment of the nucleus, site of synthesis of ribosomal RNA

Nucleus, a large structure at the center of the cell containing the bulk of the cellular DNA

Oregon R, a strain or race of fruit flies with a long life span

Organelle, small, independent particle in the cell such as the mitochondrion

Osteomalacia, soft bones, believed to be due to aluminum poisoning, boron deficiency or other reasons

Osteoporosis, bone loss disease

Oxygen, essential to life, 21% in air, but can be extremely toxic at high levels

Ozone, smog pollutant composed of three oxygen atoms and highly reactive, involved in lipid peroxidation

Photosensitizer, small molecule that increases the chemical effect of visible light

ppm, parts per million, used to express the amount of metal ions

Progeria, a rare genetic disease resulting in premature aging in children

Rad, a unit of radiation exposure

Radon Gas, radioactive gas found in most soils and rocks, can cause lung cancer

Rate of Living, a theory of aging based on the ideas of heat production and activity

Respiratory Chain, process of converting food to energy involving consumption of oxygen

Selenium, semimetal located at the active site of the enzyme glutathione peroxidase

Senescence, aging

Singlet Oxygen, same as ordinary oxygen but with electrons in a higher energy state

SOD, abbreviation for the enzyme, superoxide dismutase

Superoxide Dismutase, antioxidant enzyme that removes superoxide free radicals

Superoxide Radical, free radical composed of two oxygen atoms and one unpaired electron

Suppressor Proteins, proteins attached to the chromosomes, which determine which DNA molecules are expressed

Swedish C, a strain or race of fruit flies with a short life span

T-lymphocyte, one of the cells of the immune system

Transferrin, iron-transporting protein

Tritium, mildly radioactive form of the hydrogen atom found in all water samples

Triplet Oxygen, the ordinary oxygen that we breathe

Uric Acid, powerful antioxidant

Valence, electronic charge on an atom

Vitamin A, fat-soluble vitamin found in animal products, can be very toxic

Vitamin D, fat-soluble vitamin found in animal products, excessive intake could be toxic

Vitamin E, fat-soluble antioxidant found in all vegetable oils

Werner's Syndrome, a rare genetic disease resulting in premature aging in young adults

PUBLICATIONS, for those who want more detail.

Massie, HR and BH Zimm. 1965. Molecular weight of the DNA in the chromosomes of B. subtilis. Proc. Nat. Acad. Sci. USA, 54:1636-1641.

Massie, HR and BH Zimm. 1965. The use of hot phenol in preparing DNA. Proc. Nat. Acad. Sci., USA 54:1641-1643.

Massie, HR and BH Zimm. 1969. Kinetics of denaturation of DNA. Biopolymers 7:475-493.

Baird, MB, HV Samis and HR Massie. 1971. Recovery from zoxazolamine paralysis and metabolism in vitro of zoxazolamine in aging mice. Nature, 233:565-566.

Baird, MB, HV Samis and HR Massie. 1971. Changes in Drosophila catalase activity associated with preadult development. Dros. Info. Service, 47:81.

Erk, FC, HV Samis, MB Baird and HR Massie. 1971. A method for the establishment and maintenance of an aging colony of Drosophila. Dros. Info. Service, 47:130.

Massie, HR, HV Samis and MB Baird. 1972. The effects of the buffer HEPES on the division potential of WI-38 cells. In Vitro, 7:191-194.

Samis, HV, MB Baird and HR Massie. 1972. Renewal

of catalase activity in Drosophila. J. Insect Physiol., 18:991-1000.

Massie, HR, HV Samis and MB Baird. 1972. The kinetics of degradation of DNA and RNA by H_2O_2 Biochim. Biophys. Acta, 272:539-548.

Massie, HR, MB Baird, RJ Nicolosi and HV Samis.1972. Changes in the structure of rat liver DNA in relation to age. Arch. Biochem. Biophys., 153:736-741.

Samis, HV, HR Massie and MB Baird. 1972. Hydrogen peroxide, catalase activity, nucleic acids and senescence. In: First Rocky Mountain Symposium on Aging. HR Sobel, editor. Univ. Colorado, Fort Collins, pp.119-132.

Samis, HV, MB Baird and HR Massie.1972. Senescence and the regulation of catalase activity and the effect of hydrogen peroxide on nucleic acid. In: Molecular Genetic Mechanisms in Development and Aging, M Rockstein and G Baker, editors. pp.113-143.

Baird, MB, HV Samis, HR Massie and RJ Nicolosi. 1972. A method for the determination of catalase activity in individual Drosophila. Dros. Info. Service, 48:153.

Nicolosi, RJ, MB Baird, HR Massie and HV Samis. 1973. Cyclic AMP levels in preadult and adult D. melanogaster. Dros. Infor. Service, 49:122.

Nicolosi, RJ, MB Baird, HR Massie and HV Samis. 1973. Senescence in Drosophila II. Renewal of catalase activity in flies of different ages. Exp. Geront.,8:101-108.

Samis, HV, MB Baird and HR Massie. 1974. Deuterium oxide effect on temperature-dependent survival in populations of Drosophila melanogaster. Science, 183:427-428.

Massie, HR, MB Baird and HV Samis. 1974. Prolonged cultivation of primary chick cultures using organic buffers, In Vitro, 9:441-444.

Baird, MB, JA Zimmerman, HR Massie and HV Samis. 1974. Response of liver and kidney catalase to alpha-pchlorophenoxy isobutyrate (clofibrate) in C57BL/6J male mice of different ages. Gerontologia, 20:169-178.

Massie, HR, DS Thompson and LJ Colarusso. 1975. Discontinuous DNA replication and molecular events preceding DNA replication in Bacillus subtilis. Arch. Biochem. Biophys., 167:213-229.

Baird, MB, RJ Nicolosi, HR Massie and HV Samis. 1975. Microsomal mixed function oxidase activity and senescence I. Hexobarbital sleep time and induction of components of the hepatic microsomal enzyme system in rats of different ages. Exp.Geront.10:89-99.

Baird, MB, HV Samis, HR Massie and JA Zimmerman. 1975. A brief argument in opposition to the Orgel hypothesis. Gerontologia, 21:57-63.

Massie, HR, MB Baird and MM McMahon. 1975. Changes in the structure of Drosophila melanogaster DNA during development and aging. Mech. of Ageing and Develop. 4:113-122.

Massie, HR, MB Baird and TR Williams. 1975. Lack of increase in DNA crosslinking in Drosophila melanogaster with age. Gerontologia, 21:73-80.

Massie, HR, MB Baird and MM McMahon. 1975. Loss of mitochondrial DNA with aging in Drosophila melanogaster. Gerontologia, 21:231-238.

Baird, MB and HR Massie. 1975. A further note on the Orgel error hypothesis and senescence. Gerontologia, 21:240-243.

Baird, MB, JA Zimmerman, HR Massie and HV Samis. 1975. Response of liver and kidney catalase to alpha-p-chlorophenoxyisobutyrate (clofibrate) in C57BL/6J mice of different ages. Gerontologia, 21:240-243.

Zimmerman, JA, HV Samis, MB Baird and HR Massie. 1975. Properties of catalase molecules in rats of different ages. In: Explorations of Aging, editors, VJ Cristofalo, J

Roberts and RC Adelman, Advances in Experimental Biology and Medicine, Plenum Press, NY, pp. 291-293.

Massie, HR, MB Baird and MJ Piekielniak. 1976. Ascorbic acid and longevity in Drosophila. Exp. Geront. 11:37-41.

Massie, HR and MB Baird. 1976. Catalase levels in Drosophila and the lack of induction by hypolipidemic compounds. Mech. of Ageing and Develop. 5:39-44.

Birnbaum, LS, MB Baird and HR Massie. 1976. Pregnenolone 16 alpha-carbonitrile-inducible cytochrome P450 in rat liver. Res. Comm. in Chem. Path. and Phar. 15:553-562.

Baird, MB, HV Samis, HR Massie, JA Zimmerman and GT Sfeir. 1976. Evidence for altered hepatic catalase molecules in allylisopropylacetamide-treated mice. Biochemical Pharmacol., 25:1101-1105.

Baird, MB, LS Birnbaum, HV Samis, HR Massie and JA Zimmerman. 1976. Allylisopropylacetamide preferentially interacts with the phenobarbital-inducible form of rat hepatic microsomal P-450. Biochem. Phar. 25:2415-2417.

Baird, MB and HR Massie. 1976. Subcellular distribution of renal and hepatic catalase activity in senescent rodents. Exp. Geront., 11:167-170.

Baird, MB, JA Zimmerman, HR Massie and LV Pacilio. 1976. Microsomal mixed-function oxidase activity and senescence-II. In vivo and in vitro hepatic drug metabolism in rats of different ages following partial hepatectomy. Exp. Geront. 11:161-165.

Baird, MB, HR Massie and MJ Piekielniak. 1977. Formation of lipid peroxides in isolated rat liver microsomes by singlet molecular oxygen. Chem. Biol. Interactions, 16:145-153.

Baird, MB, LS Birnbaum and HR Massie. 1977. Presence of a high molecular weight form of catalase in enzyme purified from mouse liver. Biochem. J. 163:448-453.

Massie, HR, MB Baird and TR Williams. 1978. Increased longevity of Drosophila melanogaster with diiodomethane. Gerontology, 24:104-110.

Massie, HR and TR Williams. 1978. Invariance of longevity for Drosophila fed N,N-diethylhydroxylamine. Toxicology, 10:203-204.

Massie, HR, VR Aiello and AA Iodice.1979. Changes with age in copper and superoxide dismutase levels in brains of C57BL/6J mice. Mech. of Ageing and Develop., 10:93-99.

Massie, HR and TR Williams. 1979. Increased longevity of Drosophila melanogaster with lactic and gluconic acids. Exp. Geront. 14:109-115.

Massie, HR and VR Aiello. 1979. Changes with age in cadmium and copper levels in C57BL/6J mice. Mech. of Ageing and Develop. 11:214-225.

Massie, HR, JR Colaccico, and VR Aiello. 1980. Phenytoin-induced serum copper and ceruloplasmin in C57BL/6J mice of different ages. Age, 3:33-37.

Massie, HR and TR Williams. 1980. Singlet oxygen and aging in Drosophila. Gerontology, 26:16-21.

Massie, HR, VR Aiello and TR Williams. 1980. Change in superoxide dismutase activity and copper during development and ageing in the fruit fly, Drosophila melanogaster. Mech. Ageing Dev., 12:279-286.

Massie, HR, JR Colaccico and TR Williams. 1981. Loss of mitochondrial DNA with aging in the Swedish C strain of Drosophila melanogaster. Age, 4:42-46.

Massie, HR, VR Aiello and TR Williams. 1981. Cadmium: Temperature-dependent increase with age in Drosophila. Exp. Geront., 16:337-341.

Massie, HR, TR Williams and VR Aiello. 1981. Superoxide dismutase activity in two different wild-type strains of Drosophila melanogaster. Gerontology, 27:205-208.

Massie, HR, TR Williams and JR Colacicco. 1981. Changes in pH with age in Drosophila and the influence of buffers on longevity. Mech. Ageing Dev. 16:221-231.

Massie, HR, TR Williams and VR Aiello. 1983. Influence of dietary cadmium chelators on the survival of Drosophila. Gerontology, 29:226-232.

Massie, HR, VR Aiello and TR Williams. 1983. Chromium levels and aging in mice and Drosophila. Age, 6:62-65.

Massie, HR, VR Aiello and V. Banziger. 1983. Iron accumulation and lipid peroxidation in aging C57BL/6J mice. Exp. Geront. 18:277-285.

Massie, HR, TR Williams and VR Aiello.1984. Influence of dietary copper on the survival of Drosophila. Gerontology, 30:73-78.

Massie, HR and TR Williams. 1984. Effect of Ginseng on the life span of Drosophila. Age, 7:17, 17-18.

Massie, HR and VR Aiello. 1984. Excessive intake of copper: Influence on longevity and cadmium accumulation in mice. Mech. of Ageing and Develop., 26:195-203.

Massie, HR and VR Aiello 1984.The effect of dietary methionine on the copper content of tissues and survival of young and old mice. Exp. Gerontology, 19:393-399.

Massie, HR, VR Aiello and TJ Doherty. 1984. Dietary vitamin C improves the survival of mice. Gerontology, 30:371-375.

Massie, HR, VR Aiello and TR Williams. 1985. Iron accumulation during development and ageing of Drosophila. Mech. of Ageing and Develop., 29:215-220.

Massie, HR, TR Williams. 1985. Wisconsin ginseng and aging. Age, 8:21.

Massie, HR, TR Williams and AA Iodice. 1985. Influence of anti-inflammatory agents on the survival of Drosophila. J. Gerontology, 40:257-260.

Massie, HR, and TR Williams. 1985. Effect of sulfur-containing compounds on the life span of Drosophila. Age, 8:128-135.

Massie, HR, TR Williams and VR Aiello. 1985. Excess dietary aluminum increases Drosophila's rate of aging. Gerontology, 31:309-314.

Massie, HR, JR Ferreira, Jr. and LK DeWolfe. 1986. Effect of dietary beta-carotene on the survival of young and old mice. Gerontology, 32:189-195.

Massie, HR. 1986. Metal ions, mitochondrial DNA and aging. In: Insect Aging (K-G Collatz and RS Sohal, eds) Springer-Verlag, Berlin, Heidelberg, pp. 142-154.

Massie, HR and TR Williams. 1987. Mitochondrial DNA and life span changes in normal and dewinged Drosophila at different temperatures. Exp. Gerontology, 22:139-153.

Massie, HR and KA Kogut. 1987. Influence of age on mitochondrial enzyme levels in Drosophila. Mech. of Ageing and Develop., 38:119-126.

Massie, HR. 1988. Chemicals. In: Drosophila as a Model Organism for Ageing Studies. (FA Lints and MH Soliman, eds) Blackie, Glasgow and London, pp. 59-70.

Massie, HR, VR Aiello and RS Tuttle. 1988. Aluminum in the organs and diet of ageing C57BL/6J mice. Mech. of Ageing and Develop., 45:145-156.

Massie, HR, TR Williams and LK DeWolfe. 1989. Changes in taurine in aging fruit flies and mice. Exp. Gerontology, 24:57-65.

Massie, HR, VR Aiello, TR Williams and LK DeWolfe. 1989. Calcium and calmodulin changes with aging in Drosophila. Age, 12:7-11.

Massie, HR, VR Aiello and LK DeWolfe. 1989. Calcium

and calmodulin changes with ageing in C57BL/6J mice. Gerontology, 35:100-105.

Massie, HR, SJ Whitney, VR Aiello and SM Sternick. 1990. Changes in boron concentrations during development and ageing of Drosophila and effect of dietary boron on life span. Mech. of Ageing and Develop., 54:1-7.

Massie, HR, VR Aiello, ME Shumway and T Armstrong. 1990. Calcium, iron, copper, boron, collagen, and density changes in bone with aging in C57BL/6J male mice. Exp. Gerontology, 25:469-481.

Massie, HR and SJP Whitney. 1991. Preliminary evidence for photochemical ageing in Drosophila. Mech. of Ageing and Develop., 58:37-48.

Massie, HR, ME Shumway, SJP Whitney, SM Sternick and VR Aiello. 1991. Ascorbic acid in Drosophila and changes during aging. Exp. Gerontology, 26:487-494.

Massie, HR, VR Aiello and SM Sternick. 1991. Comparative survival of C57BL/6J mice on two commonly used mouse diets. Age, 14:53-56.

Baird,MB, HR Massie and JL Hough. 1991.Life expectancy and coronary artery disease. Circulation, 84:2607.

Massie, HR, ME Shumway and SJP Whitney. 1991. Uric acid content of Drosophila decreases with aging. Exp. Gerontology, 26:609-614.

Massie, HR, VR Aiello and SJP Whitney. 1992. Lead accumulation during aging of Drosophila and effect of dietary lead on life span. Age, 15:47-49.

Massie, HR and VR Aiello. 1992. Lead accumulation in the bones of aging male mice. Gerontology, 38:13-17.

Massie, HR and SJP Whitney. 1993. Effect of shielding from environmental neutrons on the life span of Drosophila: Preliminary evidence. Age, 16:31-34.

Sternick,SM, HR Massie and SJP Whitney. 1993. Changes

with ageing in total dolichol and dolichol fractions in Drosophila. Mech. of Ageing and Develop., 67:91-99.

Massie, HR, ME Greco and L Vadlamudi. 1993.The brain-to-liver mercury ratio increases with aging in mice. Exp. Gerontology, 28:161-167.

Massie, HR, VR Aiello and TR Williams. 1993. Inhibition of iron absorption prolongs the life span of Drosophila. Mech. Ageing and Develop., 67:227-237.

Massie, HR, SJP Whitney and VR Aiello. 1993. Probucol and aging. Age, 16:53.

Massie, HR, VR Aiello and TR Williams. 1993. Influence of photosensitizers and light on the life span of Drosophila. Mech. Ageing and Develop., 68:175-182.

Massie, HR, W Ofosu-Appiah and VR Aiello. 1993. Elevated serum copper is associated with reduced immune response in aging mice. Gerontology, 39:136-145.

Massie, HR, VR Aiello, TR Williams, MB Baird and JL Hough. 1993. Effect of Vitamin A on longevity. Exp. Gerontology, 28:601-610.

Massie, HR. 1994. Effect of dietary boron on the aging process. Environ. Health Perspectives, 102(Suppl. 7):45-48.